Also by Gianaclis Caldwell

Nonfiction

Holistic Goat Care

Mastering Artisan Cheesemaking

Mastering Basic Cheesemaking

The Small-Scale Dairy

The Small-Scale Cheese Business

Fiction

The Binding

The Useful Goat

The Ultimate Guide for Milk, Meat, Fiber, Hides, and More

Gianaclis Caldwell

Copyright © 2025 Gianaclis Caldwell

All rights reserved

Cover and interior Gianaclis Caldwell

Copyediting Kenneth Zink

Library of Congress Cataloguing-in-Publication Data available

ISBN 978-0-9861907-5-9

eISBN 978-0-9861907-4-2

Thank you for buying an authorized edition of this book. Complying with copyright laws by not reproducing, scanning, or distributing any part or version of this book without permission is of great help to hard working authors and publishers. Thank you for supporting our storytelling and sharing efforts!

Cover of photo of pack goats courtesy of Marc Warnke and packgoats.com

Grants Pass, Oregon

Foreword i
Useful Goats iii
Part I Milk 1
   Chapter One The Ideal Milk Goat 1
   Chapter Two Dairy Goat Breeds 12
   Chapter Three Registries and Breed Club Programs 23
   Chapter Four Harvesting Milk 29
   Chapter Five Cheese and Dairy Products Primer 38
Part II Meat 51
   Chapter Six The Ideal Meat Goat 53
   Chapter Seven Meat Goat Breeds 59
   Chapter Eight Harvesting Meat 64
   Chapter Nine Marketing Goat Meat 82
   Chapter Ten Meat Products Primer 93
Part III Fiber and Hides 99
   Chapter Eleven The Ideal Fiber Goat 101
   Chapter Twelve Fiber Goat Breeds 109
   Chapter Thirteen Harvesting Fiber 116
   Chapter Fourteen Harvesting Hides 123
Part IV Pack, Cart, Brush, Pet, and More 153
   Chapter Fifteen Pack and Cart Goats 155
   Chapter Sixteen The Brush Goat 171
   Chapter Seventeen Pet, Therapy, and Yoga Goats 179
   End Note 184
Resources 189
Index 195
Let's Connect 199
About the Author 201

For goats and their people and all those that have helped me along the way.

Foreword

It's been almost ten years since I set out to write a book *all about raising about goats*—the book that ultimately became *Holistic Goat Care* (published in 2017 by Chelsea Green). The original title was *The Complete Goat*. However, as is the way of publishing, it was decided the manuscript was too long, meaning it would have a cover price higher than ideal. So, I cut about 25% of the book and rewrote it all to focus on more general topics.

The seasonal pasture at Pholia Farm, 2017

I recently started thinking about all that work—and all those words! Maybe they'd be useful to folks. And then I happened to see a review of *Holistic Goat Care* that indicated the same need. So here are those missing chapters, updated and expanded. They focus on the specialization of goats for milk, meat, fiber, pack, and pet. Remember,

Foreword

this book is not a stand-alone volume about caring for goats, but a companion to *Holistic Goat Care* or whatever goat knowledge compendium you've chosen.

In 2020 our lives on the farm changed, not primarily from the pandemic, but when both mother-in-law and my mother's health took turns for the worst. We found ourselves shifting into caregiver mode and no longer able to properly continue our goat breeding and milking operation. My mother lived next door until she passed in 2023, while Vern's folks live nearby and are doing quite well, all things considered. Either way, we found ourselves contemplating the brevity of life and our own aging bodies. It was with heavy hearts that we put the farm on the market in the early spring of 2024. It sold a few months later to a lovely couple who, while they had no aspirations for making cheese, happily adopted the remaining livestock as pets.

As my life continues without goats, know that every time I hear about my books having been of help to others, I feel great satisfaction and fulfillment. I want to be with all of you in the barn or creamery, hanging out (even if I'm just the two-dimensional me on paper!).

In the back of this book, please find the latest ways to reach out and stay connected. And as always, happy goatkeeping!

Gianaclis

Useful Goats

Although some might argue that cattle are more useful than goats due to their larger size and therefore capability of providing volume and heavy draft work, I counter that the goat, for those exact reasons, wins the contest. For it is its small size and affordability that makes it more useful to the greatest number of people. To paraphrase Dr. C. Naaktgeboren says in his amazing book *The Mysterious Goat*, "What could be more noble than to provide for the poor?". Still, cattle have long been associated with wealth and goats with the impoverished. In her book, *Goats in America: A Cultural History* (due out in fall of 2025), Tami Parr states:

> *"The story of goats in nineteenth-century American cities is the story of the poor and working classes who kept goats as a means of survival in cities, the story of city authorities and their ongoing struggles to contain goats, as well as the story of the tenacity of the animals themselves, which roamed cities in search of food and were singularly adept at eluding measures to control them."*

Many immigrants arrived hoping to prosper and emulate European standards of wealth (which didn't include keeping goats). But when it comes down to finding an animal that can thrive almost anywhere—keeping their tenders alive along with themselves—goats win the prize.

Over the centuries since Europeans began taking the North American continent for their own, goats have had their ups and downs. Even in the few years since I began writing *Holistic Goat Care*, things have shifted. As with farming in general, there's been an overall drop in the number of goats in the United States (or at least in those counted by

Useful Goats

the annual USDA census). Still, it's a gentle drop, and the number of dairy goats actually increased.

I'm certain this book will garner me some scathing reviews by animal rights activists and those who ardently abstain from eating meat and using animal products. Even folks who readily wear leather and eat burgers might balk at harvesting meat from goats, much less skinning them and turning the hide into leather. I want to honor that paradigm right here! You need to do what you are comfortable with. But know this, being a healthy vegetarian (as I have been for thirty years), much less a vegan, is a privilege. Much of the world now lives with access to a variety of nutritional options, allowing for choices, and arguably, passionate viewpoints, that would not be sustainable in another time or place. No matter how you feel, know that understanding traditional, and even primitive, foodways is a way of honoring both our own past and the realities many humans still face.

Goats by the Numbers (updated from *Holistic Goat Care* page 15)

Angora goats

1997: 829,263 (37% of total goat population)

2002: 300,756 (11.9%)

2015: 150,000 (6%)

2024: 105,000 (5%)

Meat goats

1997: 1,231,762 (55%)

2002: 1,938,924 (77%)

2015: 2,100,000 (80%)

2024: 1,950,000 (79%)

Dairy goats

1997: 190,588 (8% of total)

2002: 290,789 (11.5%)

2015: 375,000 (14%)

2024: 415,000 (17%)

Source: USDA Goat and Sheep Census

Part I
Milk

Goat milk provides valuable nutrition for people around the world. Here, goats being milked at the Goat Research Station, Bandipur, Nepal. Photo by Daniel Laney.

Chapter One

The Ideal Milk Goat

A ny adult doe of *any* breed who has given birth is potentially a milk goat! By the same token, every goat bred for milk production might also be a meat goat. For this chapter, we'll focus on breeds that have been genetically selected and are marketed primarily as milk goats. We'll talk about characteristics considered ideal for all dairy goats, the main breeds used in most of the Americas and Europe, and breed clubs, registries, and production records and programs (more on those in Chapter 3).

Dairy goats at work at Toluma Farms, Tomales, California.

Part I   MILK

The ideal milk goat is not just one that looks good and milks well, but one that has a good work ethic, disposition, and, of course, health. Dairy goats are handled a great deal and should be well mannered, both for your sake and theirs. Stressed goats and/or farmers are less productive and happy! You will find people have stereotypes about what each breed of dairy goat offers regarding personality (and of course, whatever breed you love must be the best, right?). In reality, personality has as much or more to do with the traits the breeder has selected for over time. For example, a working dairy farm is more likely to say farewell to high-strung, nervous, noisy, or otherwise difficult goats than a hobby breeder is. Either way, when visiting farms, pay attention to how the animals behave and observe how easy they are to handle.

Conformation

A milk goat's conformation is assessed by looking at four categories: general appearance, dairy strength, body capacity, and mammary system. You don't have to be a licensed dairy goat judge to know what to look for, but it takes some practice to get good at *seeing and feeling* the subtle differences between an average animal and a superior one. In my old "dairy cow" days I participated several times in cow judging contests and was a dairy cattle 4-H leader, there I learned the basics of assessing dairy animals. When I got our first dairy goats, I volunteered at an American Dairy Goat Association judges training conference and learned a great deal, but the most useful training came from having our herd evaluated annually during a session called *linear appraisal* (we'll cover linear appraisal later in this section).

The Ideal Milk Goat

General Appearance

Think of this category as looking at the whole goat and how it is, quite literally, put together. General appearance looks at the underlying bone structure and alignment of the entire animal. A good dairy goat should be built to hold up well over a long career of providing milk or breeding such as with good feet and legs, shoulder blades that are snug against the body, and a head and neck that are built to eat easily and eat more, in other words, take in more feedstuffs that can be converted to milk—more milk. This category also looks at whether the goat fits the desirable characteristics of her breed. For example, if the breed should have long floppy ears, but this particular animal has ears that stick out to the side (often called *airplane ears*) then that would be an undesirable general appearance trait.

A beautiful herd of French Alpines at Rivers Edge Chèvre, Logsden, Oregon.

It is easiest to judge the general appearance of a goat as it moves about their pen—without someone trying to make it look its best as is done in the show ring. Watch how the animal moves as it walks: do its legs move forward straight from the joints or do they paddle or swing?

Part I MILK

As its feet touch the ground and bear weight, do the pasterns support that weight well, or do they seem weak? Is their back (their topline) flat and level both when walking and standing or does it dip or sway? Is their rump slightly sloped when you look at it from the side, but wide and nearly level from the rear?

With the goat standing still, take a good look at its head. There should be no over or underbite and the muzzle should be moderately broad and not narrow or pointed. (A goat's muzzle is naturally narrower in comparison to a cow's or sheep's though, with the goat muzzle designed to squeeze between branches and seek out tender leaves in shrubs—instead of moving flat to the ground to nibble on grass). Its ears should fit the description for that breed. The head should be held somewhat higher than the back, and the back should look level or even angle slightly downhill from the goat's shoulders to the tail. Now run your hands along the goat's body, feeling how its neck blends into its shoulders and how the shoulder blades sit against the body. Then run your hand down the front legs. There should be no swelling in the knees (note: quadrupeds' front legs do not have actual knees; the part we call a knee is more akin to our wrist) and the bones of the knee should round slightly above the lower leg, not arch backward. The back legs should have similar strength and alignment. In the back leg, the true knee (that atomically correlates with our own knee) is the stifle joint, tucked up against the body. Between that and the foot is the hock, which is akin to our ankle in its bone structure. When the goat stands with its feet squared beneath the body, the line between foot and hock should be straight while the line between the hock and stifle should curve forward.

The Ideal Milk Goat

Dairy Strength

This category (formerly called *dairy character*) looks at the characteristics linked to the animal that correlate to being good at producing milk—or, in the case of the buck, making babies that will be good at giving milk. A good way to become comfortable with assessing dairy strength is by comparing a meat goat or beef cow side by side with a dairy goat or dairy cow. Then the differences are much easier to see. (Note that many goats used for meat are not necessarily purposefully bred for meat, meaning they might look more like a dairy goat in general appearance. When I make comparisons in this book, it is to a more ideal, purpose-bred meat goat.)

When looking at an animal with good dairy strength, you will see a body that looks slender, but not emaciated, when compared to their meat-producing counterpart. We call this "free of excess flesh." The muscles are flat and long, rather than bulky and rounded, and there is a lack of excess body fat. Observing and feeling the thigh of the goat (the muscle between the rump and stifle joint) is a good place to assess for excess flesh. Be aware that a good dairy goat can become overweight and hide her dairy strength. This is sometimes seen when an animal is not giving milk and hasn't been bred in some time. If you suspect that a goat is simply overweight (rather than lacking in underlying dairy strength) assess the other characteristics associated with this trait, but also be suspicious that her metabolism might be more suited to making fat rather than milk.

A good dairy goat looks angular rather than blocky. This may sound similar to the qualities of excess flesh, but it is more about the shape created by their bone structure (it's evaluated separately from general appearance, though).

Part I   MILK

A classic Toggenburg doe being exhibited, after being milked-out, by breeder and judge Jennifer Tereba.

Next feel the goat's ribs and skin. There should be a decent amount of space between each rib, called *openness*; the rib bones should be flat, not rounded; and they should slope from the spine gently backwards, rather than straight down to the ground. The flatness of bone felt in the ribs can be observed in the leg bones as well—again, especially when compared to a meat-animal counterpart. The skin of a dairy animal should be soft, pliable, and "loose." The neck skin of a breeding-age buck is often covered with wrinkles from the loose skin. You can check for skin texture at the same time as feeling the ribs—just grab a handful of skin and pull gently out while you also rub it between your fingers. Luckily, the goat shouldn't mind.

The Ideal Milk Goat

Body Capacity

"Body capacity" doesn't refer to the number of people that can fit in a goat barn, but to the structure of the goat as it relates to space in her body for food and what is required to turn that food into milk. Not just capacity for food intake, but for oxygen and blood flow, in other words, the lungs and heart. A goat that is narrow of chest and rib cage will not be able to process feed as productively as one that has good body capacity. Regarding space for food intake, don't confuse body capacity with obesity or abdominal distention. We all have a goat or two who due to her age and number of births, has a barrel so well rounded that from a distance people assume she is overweight or carrying quadruplets, but there is a big difference between that (a condition called *dropped stomach*) and body capacity.

Stand in front of the goat and look at the space between her front legs, there should be ample width. If the front legs are close together at the chest, you can bet that the rest of the body will also lack body capacity. Next put your hands on the animal's back and run them down her rib cage on both sides. You should feel an ample outward arc, not a steep drop.

Mammary System

It all comes down to the udder: the make-or-break of a great dairy goat. I've had many gorgeous young does—grand champions even—who once they had their first babies and their mammary system matured, were duds—at least in the udder-department. By the same token, I've had beautiful bucks, also grand champions, that produced only daughters with disappointing udders. This is not only why it takes many years to develop a consistent herd, but what makes it difficult to choose, with any certainty, a great baby goat! When a goat is young, you can inspect their teats though, and get a good idea of their future size and placement. But the udder and the milk production are unpredictable. The best thing you can do is assess the mother's and grandmother's udders and hope for strong genetics.

Part I   MILK

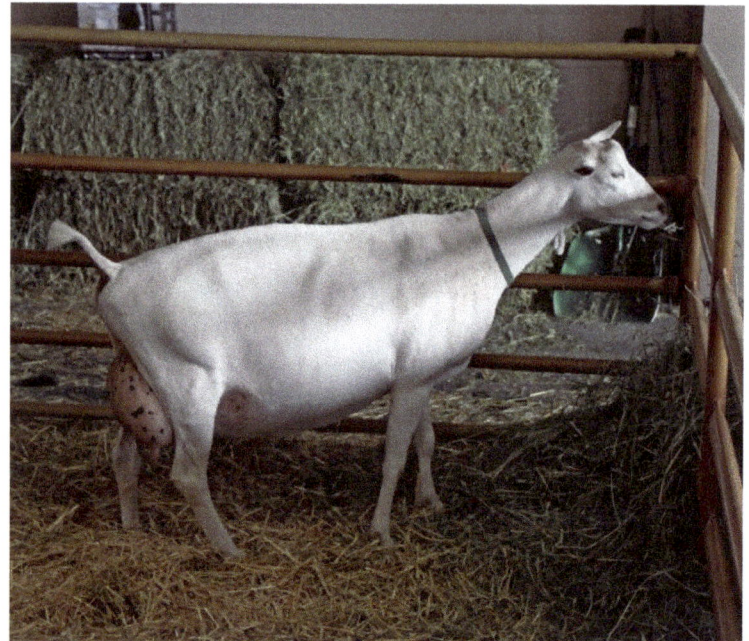

This beautiful Saanen doe doesn't need to be posed to show off her excellent general appearance, dairy strength, and mammary system.

When analyzing the mammary system of a goat in milk, it is quite helpful to know when she kidded (had her babies); her age and how many times she has kidded; and how many babies she had. Each of these factors will greatly affect how much milk she makes, and to a great degree, the size of her udder. Most goat breeds produce more milk between about 30-100 days after they gave birth. Nigerian Dwarfs (at least at this point in their development as a breed) generally peak earlier, before 60 days. If you are looking at a doe who has been in milk for 8-10 months or more, her udder will not look as "lush" and productive, even if it is allowed to fill to maximum. When the udder is past its peak, it is sometimes referred to as "stale." (Don't worry, the milk itself isn't musty, though!) Depending on her age and how many times a goat has kidded (the term is *parity*), her udder will reflect these influences.

Generally, the mammary system will look its best by the third or fourth freshening. As I mentioned earlier, the number of kids in a litter plays a role in milk production and therefore influences the size of her udder; a good doe will make enough milk to feed all her kids.

No matter what the factors, a good mammary system has certain characteristics. Where the teats are positioned (called *placement*); teat size and ease of milking; the capacity of the udder; the pliable texture of the udder; and the attachment of the udder to the body are all important things to evaluate. Let's take a look at each of these characteristics.

Teats: In order for a goat to be easily milked, either by machine or by hand, her teats should hang straight down or be slightly angled forward. If they are located a bit to the side or forward of the lowest part of the udder, then the milk will not as easily drain down into the teat. Goat teats come in a variety of shapes and sizes, unlike cow teats which are typically more uniform in length and diameter. The goat teat can be too large or too small to easily milk by hand. A nicely sized teat is not the only goal. The opening at the end of the teat (the orifice), through which you will squeeze or pull the milk, also plays a role in the ease of milking. An orifice that is too tight can make the process take twice as long and one that is too open can be prone to leaking and udder infection (*mastitis*), especially as the doe ages.

Capacity and Texture: When a doe does great in the show ring, she is often judged a second time alongside all the other winning does, but with her udder completely empty, *milked-out*. This gives the judge a chance to evaluate the texture of the udder. Sometimes a mammary system will look full of milk, but when emptied, still look a bit large. This often means that the udder is made up of unproductive tissue and therefore is deceptive in its capacity. A healthy, productive udder should be soft and rather flappy (called *collapsible*) when empty. Sometimes a doe that has just had kids will have a full-feeling udder, after emptied, due to some inflammation and possibly not having released all the milk (called holding back), especially if she is also nursing her kids. For these reasons, it's best to withhold judgement of udder texture until several weeks after kidding.

Part I   MILK

A group of handsome Nigerian Dwarf does at the National ADGA show (one from our herd just happens to be at the front right).

Attachment: If you ever get the chance to look at photos of the best dairy goats of several decades ago, you will see a vastly different ideal udder than you do today. Indeed, on working dairies in all parts of the world, milk production usually takes priority over ideal udder attachment. That being said, a highly productive *and* well-attached udder will be able to produce milk over a much longer career than a dangly, how-low-can-you-go mammary. Goats are active creatures, and a poorly attached udder will be prone to injury from being stepped on, dragged across branches, and the wear and tear of time. The udder is held up to the body not by the skin but by internal ligaments, the most important one being the one that creates the cleft down the lengthwise line of the goat's udder. The *medial suspensory ligament* (MSL) runs the entire length of the udder and is important not only for attaching the udder, but for teat placement. A goat with a weak MSL is likely to have teats that point to the sides. Lateral (side) suspensory ligaments also support the udder. A young doe should have a much higher, tighter to her body mammary system than a senior doe. If a young doe starts with her udder too large and too low (meaning the suspensory ligament is not tight) by the end of her career (which may be shorter anyway) her mammary system will likely be at risk of injury.

Work Ethic and Disposition

This may seem like an afterthought in a breeding program, but most breeders will tell you they'll happily keep a high-producing, hardworking goat over a lesser producing, high-maintenance, beauty queen. Now, I know I just spent a lot of words telling you how to judge a goat "beauty" contest, but in reality, the look and the attitude of a great milk goat go hand in hand. Disposition, like the look of a goat, is in large part genetic. You'll find the saying "like mother, like daughter" to be true more often than not. Keep this in mind when breeding for improvements in your herd. In addition, as rather intelligent animals, goats might develop "naughty" behavior when allowed. For example, if you give your goats treats or food every time they "yell" at you, well then, guess what? They will yell even more.

In conclusion, a lot goes into the recipe for making a great milk goat. If you plan on starting a herd, know that it will take years to develop one with some consistency. I encourage you to keep your early expectations less strict than your later ones! You have to start somewhere, and so does your herd.

Part I   MILK

Chapter Two

Dairy Goat Breeds

In her (fall 2025) book, *Goats in America: A Cultural History*, author Tami Parr notes that prior to the late 19th century, milk goats in North America were indistinct in breed and more scrappy-survivor than abundant milk producers. It wasn't until cow's milk began suffering issues of safety that goat's milk saw an increase in popularity. This resulted in the import of high-yielding European breeds. Since then, the most common dairy goats in the United States are the European Alpine Mountain breeds—the big white Saanens from Switzerland; the many-colored French Alpines; the taupe-with-white-points Toggenburgs, also from Switzerland; and the russet-with-black-points Oberhasli (formerly known as the Swiss Alpine). All of these breeds are famous for their high-volume milk, erect ears, and hardworking dispositions.

Wax-cone goat's milk bottles from the 1930s Toggenburg dairy of Lilian Sandburg (wife of poet Carl Sandburg) in Hendersonville, North Carolina.

Registries and Breed Club Programs

By far the most common breed used in commercial dairies in Europe and the US is the Saanen. Alpines come in a close second. The least common is the Oberhasli, with the Toggenburg not far behind. In the US, a "new" breed has been added, the Sable. While not a new breed in terms of genetics, the Sable is from Saanen genetics, but is a color other than white.

Other popular and productive dairy goat breeds recognized by the American Dairy Goat Association include: the Roman-nosed, floppy-eared Nubians; the tiny-eared Lamanchas; the diminutive Nigerian Dwarfs; and the golden, flowing coated Guernsey goats.

Growing in popularity are a group of so-called miniature breeds, the results of crosses between Nigerian Dwarfs and the bigger breed that is being "miniaturized."

Two other breeds, not recognized by ADGA or AGS, but with their own registries, are Kinders and the very new Nigoras (Nigh-gor-uhs). Both are dual-purpose breeds, with Kinders being for milk and meat and Nigoras for milk and fiber. I'll cover Nigoras in the fiber chapter.

Here's a brief overview of dairy goat breeds in the United States. In the reference section at the back of the book, you'll find links to the different breed associations where you can learn more and maybe even connect with a breeder near you.

Note: Most breeds registered with ADGA have two herdbooks: one for purebreds and one for Americans. In a nutshell, inclusion in the American herdbook is attained after several generations of "breeding up," during which, several stages must be met by the goat to attain the breed standards.

Part I   MILK

Alpine – French (Purebred), American, and British

Origin: The Alpine areas of Europe, especially France and the Sundgau region along the eastern edge of France and the border of Switzerland. Imported to England in the early 1900s from Paris and to the United States and Canada first in 1904 and then in 1922.

Ideal Size Standards for US: Does no less than 30 inches (76 cm) at the withers and 135 pounds (61 kg). Bucks no less than 32 inches (81 cm) at the withers and 170 pounds (77 kg).

Distinguishing Characteristics: Color patterns in US: cou blanc (white neck with a dark back half and dark facial stripes and legs), cou clair (light-colored neck with a dark back half and dark facial stripes), cou noir (dark neck with a light back half and light facial stripes), sund gau (dark body and legs and light facial stripes, belly, and lower legs; this is the only color pattern for British Alpines), chamoisee (brown or reddish body with a dark face, belly, dorsal stripe, and legs), two-tone chamoisee (same as previous but with a light front half), broken (any of the above but with a white pattern overlaying the main color pattern). All white or Toggenburg color patterns are undesirable. Straight nose and erect ears.

Other: French Alpines are purebreds that can be pedigree traced back to the original imports from France. American Alpines are those bred up through the ADGA program. (See photo on page 3)

Saanen – Purebred, American, and British

Origin: Switzerland, brought to the United States and Canada in two major imports in 1904 and 1922.

Size: Does no less than 30 inches (76 cm) at the withers and 135 pounds (61 kg). Bucks no less than 32 inches (81 cm) at the withers and 170 pounds (77 kg).

Distinguishing Characteristics: White to cream colored, with white preferred. The largest of the dairy goat breeds with the most milk and the lowest in butterfat and protein. (See picture page 7)

Registries and Breed Club Programs

Toggenburg – Purebred, American, and British

Origin: Developed in the Toggenburg valley in Switzerland and imported into England in the late 1800s and into the US in the 1920s.

Size: Does no less than 26 inches (66 cm) and 120 pounds (55 kg). Bucks no less than 28 inches (71 cm) and 150 pounds (68 kg).

Distinguishing Characteristics: Shades of brown, tan, and taupe with white facial stripes, legs, and rump markings (also called *Swiss markings*). (See picture page 15)

Oberhasli

Origin: Known as Swiss Alpines until 1977 when they were renamed Oberhasli and classified by the American Dairy Goat Association as a separate breed. Originally from Switzerland. Imported to the US in the early 20th century, but breeders lost track of those imports. More imports from Switzerland in the 1930s established the breed.

Ideal Size Standards: Does no less than 28 inches (71 cm) and 120 pounds (55 kg). Bucks no less than 30 inches (76 cm) and 150 pounds (68 kg).

Distinguishing Characteristics: Reddish brown or bay color, with black legs, face, belly, and dorsal stripe. Solid black does acceptable, but not bucks.

Sable

Origin: Recognized as a separate breed by the American Dairy Goat Association in 2005. Rather than a new breed, however, Sables are from the same original genetics as Saanen goats, but are not colored white or cream (disqualifying them from the Saanen registry).

Ideal Size Standards: Does no less than 30 inches (76 cm) at the withers and 135 pounds (61 kg). Bucks no less than 32 inches (81 cm) at the withers and 170 pounds (77 kg).

Distinguishing Characteristics: Same qualities as Saanen goats but all colors are acceptable.

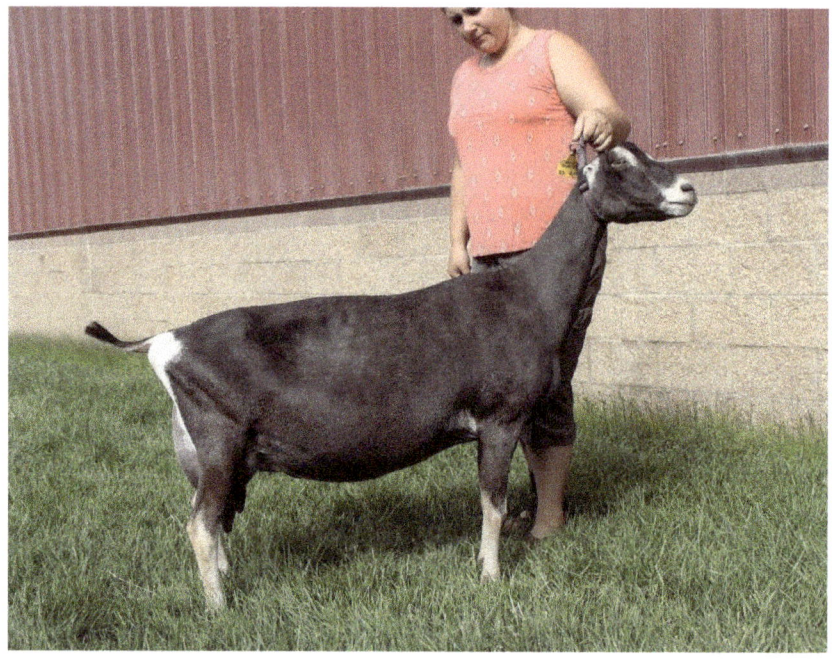

A stylish and powerful Lamancha doe (milked out) shown by Mariah Acton.

Lamancha

Origin: Developed in the United States from naturally short-eared goats thought to have Spanish ancestry. Recognized as a breed in 1958.

Ideal Size Standards: Does no less than 28 inches (71 cm) at the withers and 130 pounds (59 kg). Bucks no less than 30 inches (76 cm) at the withers and 160 pounds (72 kg).

Distinguishing Characteristics: Short ears. The shortest are called *gopher ears* and must be the type present on bucks. Elf ears, a bit longer, are acceptable on does. All coat colors are allowed.

Registries and Breed Club Programs

Nubian and Anglo-Nubian

Origin: The term "Nubian" was used to describe any goats from the Near, Middle or Far East that had the typical appearance (long, floppy ears and arched noses) of the Jumna Pari breed from India and the Zaraibi breed from Egypt. These goats were mated with goats already in England, and the resulting crossbred animals gained the name Anglo-Nubian in 1893. The first were imported to the US in the late 1890s, but what became of them was not recorded. Documented goats were imported in the early 1900s and registration of purebred Nubians began in 1917. Although an American herdbook exists, purebreds continue to be the most popular Nubians.

Size: Does no less than 30 inches (76 cm) at the withers and 135 pounds (61 cm). Bucks no less than 32 inches (81 cm) at the withers and 170 pounds (77 kg).

Distinguishing Characteristics: Long, pendulous ears (although shorter than those of the Jumna Pari whose ears reach halfway down their necks) extending at least one inch beyond the muzzle and an arched, or Roman, nose. In general, Nubians are heavy-bodied goats and are known for the high butterfat content of their milk. (See photo on page 80)

Part I   MILK

Nubians and G6S Genetic Abnormality

Nubian goats and Nubian crosses (including meat goats, such as Boers, crossed with Nubians) can carry a recessive gene that causes the animal to be deficient in a critical enzyme. The condition leads to delayed motor development, growth retardation, and early death. For that reason, animals with Nubian genetics should be tested by their breeders for the gene G6-S. The gene is believed to be present in about 25% of Nubian and Nubian crosses. When a goat inherits two genes for the G6-S mutation (one from each asymptomatic carrier parent) it is a death sentence. The kid fails to grow and thrive and usually dies before their first birthday. Goats that only inherit one gene are asymptomatic carriers.

There are many genetic defects in goats that can't be anticipated and managed through genetic testing, but fortunately, this one can. Breeders who are aware of their goats' statuses as carriers or noncarriers can make matches that limit the spread and can then test the offspring—culling as needed to further limit the gene's presence in the breed. Testing runs from $25-$30 per sample, but will increase the value of the herd and the confidence of buyers. For more on testing visit The University of California Davis website and search for G6-S-goat testing.

*SGCH Pholia Farm HB Angelica 6*M at the 2010 National Show, the first where Nigerian Dwarfs were allowed to compete. My daughter Amelia and I drove from Oregon to Louisville, Kentucky to participate*

Nigerian Dwarf

Origin: Imports of small goats from the west coast of Africa for zoos and animal parks are believed to be the origin of both the Pygmy goat and Nigerian Dwarf. The name was coined in the US by an early breeder and, for better or worse, stuck. First recognized by the American Goat Society in the 1980s and by the American Dairy Goat Association in 2005.

Size: Does no less than 17 inches (43 cm) at the withers and no more than 22.5 inches (57 cm). Bucks no less than 17 inches (43 cm) at the withers and no more than 23.5 inches (60 cm).

Distinguishing Characteristics: The smallest dairy breed. All colors acceptable. Blue eyes acceptable. Known for high-butterfat and high-protein milk.

Part I   MILK

Guernsey and Golden Guernsey (British)

Origin: The newest breed accepted into ADGA, this midsized breed is built on the genetics of the English Channel Island breed, the Golden Guernsey. The breed is still being developed by using purebred genetics and crossing them with goats with similar breed characteristics.

Size: Does no less than 26 inches (66 cm) at the withers and 120 pounds (55 kg). Bucks no less than 28 inches (71 cm) at the withers and 150 pounds (68 kg).

Distinguishing Characteristics: Gold to light blonde coat, often long-haired. Fine boned.

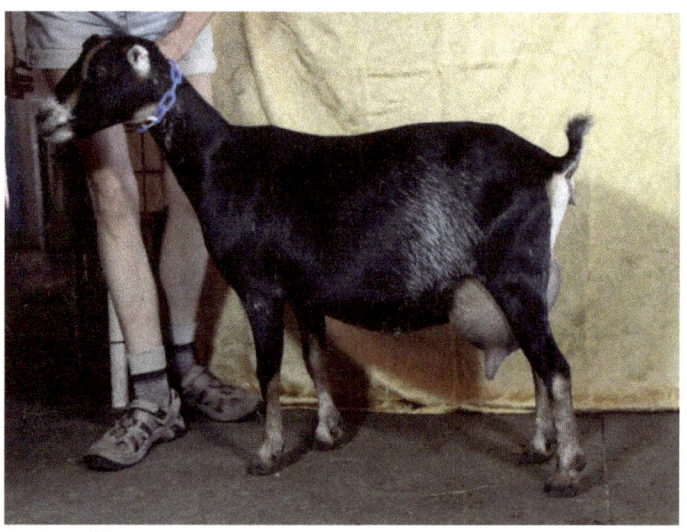

Prudence, a Nigerian Dwarf/Lamancha cross (we called them Lagerians) at Pholia Farm.

Miniature Breeds

Origin: Originally crosses between Nigerian Dwarfs and standard-sized breeds that are smaller than their full-size counterparts but retain that breed's characteristics. Miniature breeds are currently not recognized by ADGA but have their own registry.

Registries and Breed Club Programs

Size: Varies according to breed. Generally, 29-31 inches (73-79 cm) at the withers.

Distinguishing Characteristics: Typically follows the breed standards set for the full-sized breed after which it is modeled, with exception of height.

Kinder

Origin: First developed in 1986 with the cross of a Pygmy buck and Nubian does, this dual-use goat produces ample milk and meat in a compact versatile-sized goat.

Size: Does 20-26 inches (51-66 cm) at the withers. Bucks maximum 28 inches (71 cm) at the withers. Weight 100-125 pounds (45-57 kg).

Distinguishing Characteristics: Long ears, held at a gentle right angle—often called *airplane ears*. Heavier body type. All colors. (See photo on page 56.)

The True Goodness of Goat's Milk and the Myths

Years ago, I was inspired to write my first blog post after reading, for the umpteenth time, the *Journal of American Medicine*, JAMA, quoted as saying: "goats' milk is the most complete food known."

I went to the JAMA website and did a search of their archives. I found a letter to the editor in Volume 120, #4, published in September of 1942, that asked if the following quote (which the letter's author had read in another publication) was true: "The Journal of American Medical Society states that goat milk is the purest, most healthful, and most complete food known." The editor of JAMA responded that no such statement could be found in →

Part I   MILK

← JAMA. That was over eighty years ago, and people are still taking a question and turning it into a statement that the cited source never said.

This doesn't mean that goat's milk isn't amazing and wonderful! In fact, I think it's so wonderful that we don't need to rely on inaccurate hype to promote it. But here are some of the ways goat's milk is different.

The fat globules in goat's milk are easier for most people to digest, being smaller and with a weaker milk-fat globule membrane, the "wrapper" holding the fats together, making them are easier to break down and digest.

Milk Protein: Most goat's milk lacks, or is low in, he type of protein in milk that people are more likely be allergic to.

People who have trouble with lactose (milk sugar) are not allergic to it, they are intolerant: their bodies lack the enzyme needed to break milk sugar into easier-to-digest sugars. Goat's milk is NOT lactose free, or even much lower in lactose than cow's milk. However, as much of the goat's milk consumed is raw (unpasteurized or otherwise heat-treated), the enzyme (lactase) needed is often provided by natural raw milk bacteria.

You'll hear some folks claim, "Goat's milk is a better source of vitamin B than cow's milk." Here again, there is misinformation. Strictly speaking, this is correct, but the truth is that goats convert the beta carotene they consume into vitamin B, while cows do not. Fortunately, our bodies easily take the beta carotene in cow's milk and convert it into vitamin B.

The main takeaway is that when harvested properly goat's milk is nutritious and delicious.

Chapter Three

Registries and Breed Club Programs

Goat registry associations keep track of many things for the breeders who register their animals with them. Primarily they maintain what are called *herdbooks*—pedigree records for each breed and every animal of that breed born, provided they are registered or recorded with the association. This information is valuable for those wishing to track the family tree of their goats. It also provides proof of age and identification for those purchasing goats. The largest registry, the American Dairy Goat Association, or ADGA (pronounced "Ad-guh"), also collects and translates the data from production programs, such as milk production records and genetic evaluations, to help producers make breeding decisions in hopes of improving their herd. Having a goat registered or recorded with an organization also allows it to be shown in official goat shows that are sanctioned by each organization. Major registries in the United States and England include The American Dairy Goat Association (Purebreds and Americans), The American Goat Society (purebreds only and now mostly comprised of Nigerian Dwarfs), and The British Goat Society.

Breed clubs are organizations that seek to support the goals of their members. Sometimes, due to differences in just what those goals should be, there is more than one club for the same breed! You do not have to be a member of a breed club to own and register goats of that breed. Breed clubs are not an official part of the main registries mentioned above, but work in conjunction with each registry to decide and change things such as the official breed standards.

Part I   MILK

The wall inside the Pholia Farm barn with a collection breed club, ANDDA, ribbons won at shows.

Pedigree Performance Records

There are many programs and designations of awards that are included on a dairy goat's pedigree. They are usually indicated by letters and symbols placed before and/or after the animal's name. The registration paper will be updated whenever it is sent to the main office, such as when an animal is sold and transferred to the new owner. In other words, if an owner claims information and you don't see it on the pedigree, that doesn't necessarily mean it doesn't exist.

Don't worry too much if you don't readily understand all the letters, symbols, and numbers appearing on a pedigree. There's a lot to learn, especially if the goat comes from a herd involved in all of the programs possible! Let's go over the most common of these awards and designations.

Milking Awards

Dairy Herd Improvement (DHI) is a nationwide (United States) program originally designed to help cow dairy farmers improve their milk production and genetics. It was first adopted by the dairy goat world in 1939. It is not a government program, but the data collected *is* shared with the United States Department of Agriculture (USDA) and is available for all to see through a currently less-than-intuitive website at adgagenetics.org. When enrolled in DHI, producers typically call it

Registries and Breed Club Programs

"being on milk test." Know that you do not have to have registered or purebred animals to be on milk test. You don't even have to be a member of one of the goat registries! For those that do have animals recorded or registered with ADGA and AGS though, there are numerous awards and designations that can be earned by your dairy goats. (Due to its popularity and so as not to add further confusion, I'll only be covering ADGA's awards here.)

Each breed for which ADGA maintains a herdbook also has a set of criteria by which that breed can earn *milking stars*. The criteria is based on the amount of milk, milk fat, or protein produced by the animal. Depending on the age of the animal, the minimum amount that she must produce to earn her star varies. As you might guess, the minimums increase with age, until the animal passes her prime. The amounts needed to earn a star vary by breed. You can find this information in the ADGA member guidebook or on their website at adga.org.

A star is only earned once and cannot be taken away or lost (unless fraud was involved, that is). A doe who earns her star, but whose mother did not earn one, receives the designation 1*M. (The asterisk stands for the word "star.") If her daughter earns a star, she receives the designation 2*M. And then if her daughter earns a star, she is a 3*M, and so on. The higher the number you see in front of the *M, the more females in a family line have milked well enough to earn their award. So, it is a good indicator of those genetics being good at producing milk. The chain can be easily broken though, either by an owner that doesn't do milk test or if a doe dies before she can continue the line of milking stars. As with everything related to a goat's pedigree, there's much to take into account!

Bucks can also have milking stars (*B) and any number of plusses (+B) after their names. How, you may ask, can a buck—who cannot give milk—get a milking star? There are two ways. First, he can inherit it. If his mother has her star and his father has his, then he will automatically be a *B animal. It will appear after his name on his registration papers, even if he is still a kid. But how did that "first" buck get his? Or how does a buck whose parents didn't have stars earn one? When a buck produces (fathers) three daughters from three different mothers and each of these daughters earns their stars, then daddy gets his too. If he already has his *B designation then he gets his first +, so his milking star will look like this: +B. Wait, there's more. If three more daughters, again from three different mothers, get their stars, then the dad gets yet another +, and that acquisition of plusses can continue even after he passes away.

Superior Genetics

This designation appears in front of the animal's name as SG. The SG award is a bit difficult to understand and explain. I turned to ADGA's former performance chairman Lisa Shepard who summed it up like this:

> "The ADGA SG program recognizes animals in the top 15% for their breed at least once during the life of the animal according to their Production/Type Index (PTI) value. This value is based on complex genetic evaluations for both milk production and type (conformation) which are then combined to calculate the PTI ranking. The Index values are computed twice a year and come out in two forms: one which emphasizes production and one that emphasizes type. Once earned, the SG designation becomes a permanent part of the animal's record."

I'll admit, I still don't quite get it—and I had many goats with the award! Still, it is a helpful tool when choosing who stays in your herd and a valuable tool for those shopping for good dairy goat genetics.

Championships

Show-ring awards, and how they are accumulated to create a *finished champion*, follow a very complicated set of rules. I won't worry about bogging you down with those right now! But if a doe or buck does win the crown jewel of show-goat prizes, the designation CH appears in front of their name. If they also have the superior genetics award, it will be combined and appear as SGCH.

Registries and Breed Club Programs

Linear Appraisal

Linear appraisal (LA) is another complicated but very valuable program for those seeking to better understand the strengths and weaknesses of their own herd and breed. The program collects and disseminates data meant to assist breeders in choosing sires with powerful transmitting abilities of desired traits. Unlike milk production awards, linear appraisal scores and data on individual goats participating in the program are re-evaluated periodically, usually every year. The score itself will not appear on the registration papers, but the date of the appraisal will.

Breeders typically put LA scores on their herd promotional literature. This score, a number, will appear with a set of letters (4 if it's a doe, and 3 if it's a buck). For example, a milking doe's LA score might read EAEV 85 and a buck's might read EEV 90. Each of these four letters is a score reflecting the appraiser's conclusion regarding the goat's *major category traits:* general appearance, dairy strength, body capacity, and mammary system. (See the sidebar for more on these letter scores and how they compare to a dairy goat ideal). The final score—the 85 and 90 in our examples earlier—is the number that sums up the scores as a whole.

In truth, the linear appraisal system has three parts: An evaluation of **15 "linear" traits** on a scale of 1-50, with each end of the scale representing an extreme, rather than a "good" or "bad" (ADGA has an entire booklet about it!); evaluation of the animal's **8 structural/ functional traits** with a letter/symbol score (see sidebar); and the **four major categories** (mentioned in the previous paragraph) scored with a letter/symbol and then the assignment of a **final score**.

I highly encourage any new breeder to participate in a linear appraisal session as soon as they can. Even if you are not able to bring your own animals, find a friendly breeder that will let you observe their LA session. In most ways it's more informative and more objective than any goat show or show awards. And of course, despite me saying" more objective," the program still relies upon humans; you will definitely see

Part I   MILK

that results will vary by appraiser. But over time and repeated sessions, a quite balanced picture of your herd's quality will appear.

For more on the program, visit: https://adga.org/performance-programs/linear-appraisal/

Linear Appraisal Descriptors and Percentages

Excellent **(E)** at least 90% of ideal

Very Good **(V)** 85% to 89% of ideal

Good Plus **(+)** 80% to 84% of ideal

Acceptable **(A)** 70% to 79% of ideal

Fair **(F)** 60% to 69% of ideal

Poor **(P)** less than 60% of ideal

Chapter Four

Harvesting Milk

In this chapter let's talk about harvesting delicious and nutritious goat's milk for drinking and making dairy products. If you want to know more about the needs of the goat during breeding, pregnancy, and delivery (all the stages it takes to get to the milk part) please consider referring to Chapter 8 of *Holistic Goat Care*. On page 21 earlier in this book, you'll find some facts and myths about goat milk.

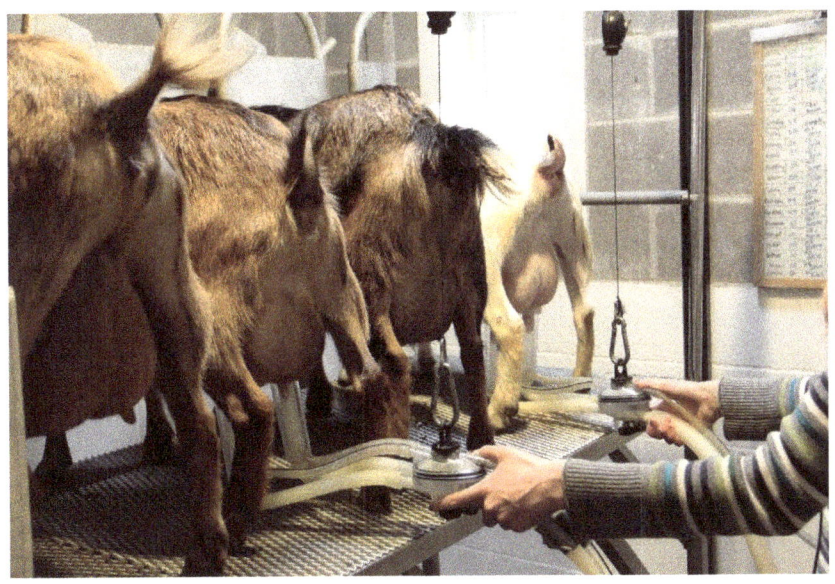

Milking at Pholia Farm

Part I MILK

Commercial Versus Home Production

There's quite a difference between producing milk strictly for home use and producing it for commercial sale. In almost every state in the United States and in many parts of the world, milk and dairy products cannot legally be sold or shared with those outside your immediate family without regulatory oversight. If you are interested in having a commercial dairy, know that almost no other food industry is as regulated as is the dairy industry. The decision to become a licensed dairy (making and selling fluid milk) or products producer (making and selling cheese, butter, or yogurt) brings with it the need for a deep understanding of regulations and requirements. I've dedicated two previous books to this topic if a deeper dive is of interest to you.

There's more than one way to milk a goat! A historical illustration of a patient goat being used to nourish an infant.

Harvesting Milk

Milking Basics

High-quality goat's milk comes only from healthy, well-nourished, and unstressed goats. The task of milking must also be done properly in order to preserve the quality of the milk. There is a saying that I like to repeat that should help you to understand how tricky maintaining that quality is, "Milk was never meant to see the light of day." Think about it: nature designed milk, from all mammals, to go directly into the baby's mouth and then stomach. Every step we humans add in our efforts to preserve it for later use adds a layer of reduction in quality and even possibly makes it unsafe. Here are the keys to ensuring you have the best possible milk supply. Please remember that the licensed and inspected facility will have a much longer checklist of requirements. For a deeper dive into all things milk-production related, please consider my book *The Small-Scale Dairy*.

1. Healthy goats that are tested for every known disease passed in milk to humans.
2. A clean place to milk that's separate from exposure (particularly during milking) to dust from pens, manure piles, and feed; is easy to keep clean and is kept clean; is well lit; and has hot and cold running water. (Inspected facilities will have a more extensive list of requirements.)
3. Milking equipment that is in excellent condition and can be kept immaculately clean.
4. Clean animals entering the milking and haven't shared barn space with pigs or poultry.
5. Healthy and clean people who do the milking.
6. The ability to quickly chill the milk to the proper temperature and keep it cold.

Part I   MILK

Milking Preparation

Fortunately for those of us wanting to milk goats, the process is helped along by the fact that goats are naturally fairly clean, especially when compared to cows. Their dislike of water and the fact that their manure is dry means that goats come into the milking parlor likely cleaner than their dairy cow cousins. This means that there is less work preparing them for milking and less likelihood of milk contamination with manure.

To further increase the ease of collecting clean milk, you can clip the hair from the goat's udder, belly, and flanks—called a *dairy clip*. The dairy clip should be updated periodically and goes a long way toward increasing the cleanliness of the milk by keeping dust and debris from entering the milking pail or being drawn in through the milking machine's air vents. (A milking machine system is not closed! It suctions a tiny bit of air into every line in order to not create a vacuum where milk would not flow readily away from the animal.)

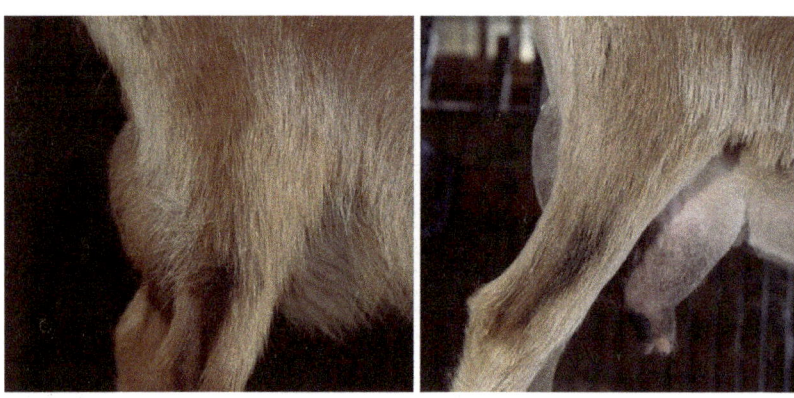

Udderly fuzzy before a dairy clip!

Once in the parlor, the udder should be lightly cleaned with a moist, not wet, cloth. A soapy solution can be used. It's critical that it not be dripping wet, as any moisture running down the udder could drip into the pail or be drawn into the milking machine teat attachments. The soapy solution is meant to clean away any debris and the rubbing will stimulate

Harvesting Milk

the letdown response of the doe so that she relaxes and releases her milk more readily.

Once letdown has begun, the first milk, called the *foremilk*, is removed by milking three to four squirts into a bowl or specially designed cup, called a *strip cup*. This helps flush out contaminating bacteria that will have invaded the teat during the interval between milking. Observe the milk for abnormalities such as clumps, blood, or mineral crystals that could indicate an udder infection (mastitis). Every month, or more frequently, the foremilk should be checked for somatic cell count (skin and white blood cells) by using an on-farm test. Some of these cells are always present, but a spike or chronically high number is indicative of a problem. I like the California Mastitis Test (CMT). It's simple, inexpensive, and effective. By doing this type of test regularly, you are likely to catch mild udder infections while they are still treatable without needing to use antibiotics. See the section on mastitis in *Holistic Goat Care* for more.

> **The Changing Paradigms of Dairy Sanitation**
>
> A common practice is to pre- and/or post-dip teats in a sanitizing solution, but recent data is showing that with a healthy udder this is not only unnecessary, but destroys protective skin flora on the teat. It's still used a lot though, particularly in high-turnover dairy parlors where workers can only take seconds to prep an udder. At home and in small dairies, more time is usually available to make up for what the chemicals can do with thorough cleaning by hand.

After the foremilk has been removed the teats can be dried with individual paper or washable cloth towels. It's always important to milk with dry hands and dry teats—even when using a machine. When either is wet, the likelihood of spreading bacteria from animal to animal or into the milk goes up. When a machine is used, the wet teat can cause the milking attachment to slip and irritate the teat or allow milk to be pushed up into the teat, leading to trauma and possible infection.

Milking Technique

Hand milking is a skill that takes a few days to master, but once you do, you will gradually increase your speed and skill to equal that of a milking machine. A milking machine is a good choice when many goats must be milked and, when equipped properly, allows for multiple goats to be milked at once. The cleaning and maintenance of a milking machine takes time and an understanding of how to use the right cleaners and sanitizing solutions. (If you are considering setting up a system for a commercial dairy, in-depth information on that topic is covered in *The Small-Scale Dairy*.) There are some inexpensive setups that look a bit like a hand-pumped milking machine and promise ease of milking, but the physics of the way most of these work on the teat are often not good for long-term use as they don't allow for good blood circulation in the teat.

Hand milking traps milk in the teat using your thumb and first finger and then gently pushes it out of the teat orifice—the opening at the bottom—with your other fingers. Machine milking uses vacuum pressure turned on and off by a unit called a *pulsator* to remove milk. The vacuum mimics how the kid nurses a bit more than hand milking does, so when the vacuum is set to the right suction level and the attachments fit well, it is quite comfortable for the goat.

Milking Equipment and Cleaning

Milk should be collected in a stainless-steel pail or bowl. As soon as the milking is completed, the milk should be poured through a milk filter (usually a disposable paper mesh, but sanitized cloth of a tight weave can also be used) and into thoroughly cleaned glass jars. If the milk cools too

Harvesting Milk

much before filtering, the fat globules in higher-fat goat's milk might solidify and clog the filter. After pouring the milk through the filter, it should be inspected for abnormalities in the milk—the same as those you look for in the strip cup.

All milking equipment and storage containers should be kept immaculately clean. Immediately after milking, rinse all equipment with lukewarm water, then wash and scrub in hot soapy water and rinse well and set to air dry. An automatic dishwasher does a fine job cleaning hand milking equipment. Just before the next use of the equipment, it's a good idea to sanitize all equipment and containers with either boiling hot water or a properly diluted sanitizer. Bleach is commonly used, but is very hard on stainless steel, causing pitting if it's too strong. The usual recommendations, including those recommended by household bleach companies, create a solution much stronger than needed, but err on the side of caution. If you want a more proper dilution, you can buy chlorine test strips to verify the strength. Our dairy inspector said that a proper dilution smells reminiscent of the scent of a chlorinated swimming pool. When properly diluted, you don't need to rinse off the sanitizer.

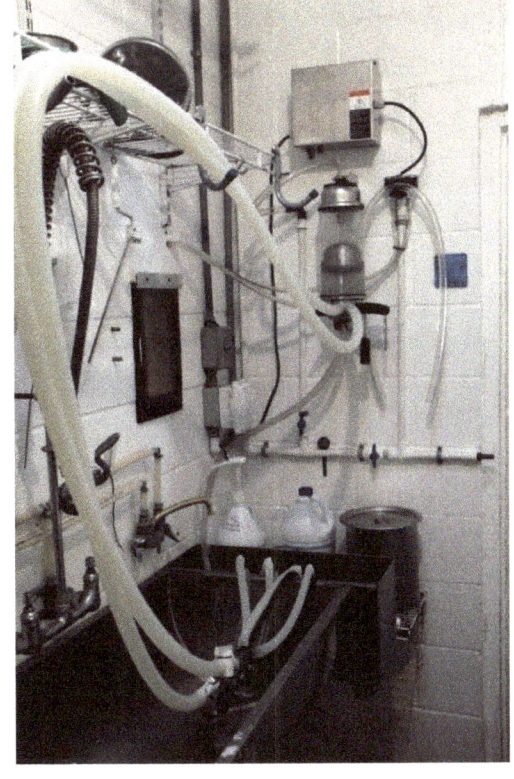

Cleaning milking machine equipment is an elaborate and extensive step in milk quality. Here, the Pholia Farm portable milking lines are connected to a "bucket-washer" which pulls and pushes the three stages of cleaning solutions through the lines.

Part I MILK

Cooling and Storing

Quickly cooling milk from the temperature at which it leaves the goat, from right about 100 F (38 C) to refrigeration temperature, at or below 40 F (4.4 C), within two hours after milking is finished is critical for milk safety and flavor. Goat's milk is particularly prone to flavor changes that leave it smelling and tasting a bit like the aroma of a buck. When it is very clean and chilled quickly the flavor should be crisp and clean, with no barnyard or buck flavors. Setting the jar in the fridge isn't an effective way to chill the milk. Small jars, under 1/2 gallon (500 ml), can be chilled in the freezer. Shake it once or twice during chilling to help it chill evenly. You can also set jars in a sink or tub of ice water. Be sure to stir or swirl the milk as it chills. When you first start figuring out how to chill your milk, verify that the temperature goal is achieved by checking it with a thermometer.

Commercial dairies have the same chilling goal, but usually accomplish it by pouring the milk or pumping it into a chilling unit called a *bulk tank*. The bulk tank stirs and chills the milk as well as records its temperature for the dairy inspector and dairy farmer. Like milking machine equipment, chilling equipment requires a great deal of cleaning and knowledge about maintenance.

Can You Freeze Goat's Milk?

When foods reach freezing temperature, the water molecules begin to form into crystals. The longer it takes for the substance to become completely frozen, the larger these crystals grow. As they grow, they damage solids around them. Take for example fresh fish: when frozen at home, then thawed, the flesh is often mushy from the resulting damage of ice crystals. Milk is no different. The solids, the milk-fat globule membrane and the proteins, are damaged during freezing. That being said, the faster milk can be brought to 0 F (18 C), the less damage will be done. In addition, "low-water" milk (basically milk that is naturally higher in fat and protein like Nigerian Dwarf and Nubian milk) has a better shot at successful freezing simply because there is less water.

To accomplish proper freezing at home without commercial flash-freezing equipment, pour the milk into a shallow container (a freezer bag works well), lay on a tray, and place at the lowest available level of a chest freezer set to its lowest temperature. Although milk can stay frozen for long periods of time, the less time it spends this way, the less damage will occur.

When you thaw the milk, if you find it clumpy or gritty, then you know damage has occurred.

Part II   Meat

Chapter Five

Cheese and Dairy Products Primer

Having written a few volumes on the topic of making cheese and dairy products, I know there is no way to adequately explore it all here. However, I'd love to give you an encouraging overview of the amazing range of dairy products you can make with goat's milk. If you're considering selling cheese commercially, remember that it's a highly regulated *manufactured* product, meaning steps have been taken that convert a commodity (milk) into a product. Even states that allow for direct, unregulated farm sales of milk are unlikely to allow the same for cheese, since the extra processing steps increase opportunities for introducing hazards. If you're interested in more on this, I encourage you to read a couple of my other books (listed in the back) on the topic.

Me in 2012 with some of Pholia Farm's raw-milk cheeses.

The Ideal Meat Goat

Goat's Milk Differences

Goat's milk offers some specific differences from cow's milk when it comes to making products. These differences are not only species related; goat's milk varies more than cow's milk due to breed and season. In addition, the lack of a specific protein (present in cow's milk) that causes fat globules to cling to each other prevents the easy separation of cream from the milk. For cheesemaking and yogurt-type products, this isn't much of an issue, but if you're wanting to make butter or skimmed milk cheeses, it's an impediment.

Species-wise, in general, goats produce a lower percentage of cheese proteins (which affects the yield of product, or in other words, how much cheese you can get from each gallon of milk) than do cows. This isn't true across the board but needs to be mentioned. When making products, this difference can create a weaker curd and lower yield, requiring gentler handling during "the make" and different expectations regarding how much curd is produced.

Regarding breed differences, Alpine, Toggenburg, Saanen, and Oberhasli goats (those high-producing Alpine breeds we mentioned earlier) have little to none of the best type of protein for making rennet-coagulated cheeses (think firm, aged styles), while Nubian, Lamancha, Nigerian Dwarf, Pygmy, and Boer goats (yes, some people milk them) do. (There are many other breeds in the world that do also, but here, I'll stick with the better-known breeds in North America—sorry!) The first group, those Alpine Mountain breeds, have milk best suited to superior soft-fresh and soft-ripened cheeses, while the other breeds' milk has a greater range of possibilities.

Lastly, there's the lack of significant "creaming," or the rise of milk-fat globules that occurs in cow's milk. With time, some cream does separate and can be hand-skimmed and saved up for butter. The alternative is a mechanical cream separator. These devices are effective, but expensive and a chore to clean. And here's the thing: goat's milk butter does NOT taste like cow's milk butter. It's delicious, but lacks the same flavor compounds most of us associate with butter.

Part II   Meat

Cheese

Cheese is any curdled milk that is drained of the resulting liquid (whey). There are three foundational "technologies" for making cheese. I'm going to describe these make-processes in a nutshell and give you a couple of get-started recipes to hopefully launch your cheesemaking experience. Know that there are many cheese varieties that are hybrids of the following three technologies!

High-Heat, Added-Acid Curd

This is the fastest and easiest way to set curd. Milk is heated, a food-grade acid (vinegar, lemon juice, etc.) is added, the milk curdles, the curd is drained, and it's done. Examples are ricotta, paneer, and queso blanco (one version). All of these can be stored frozen, but cannot age well due to their high moisture content. Paneer is my go-to cheese for ease of make and versatility! You can season it, fry it, grill it, and add it to soups and stews, as it is a non-melting cheese. The whey from cheeses in this category has little to no nutritional value, but can be used to water acid-loving plants.

Cultured Long-Set Lactic-Acid Curd

This approach takes a bit longer, but, like making yogurt, it's straightforward and hard to mess up. High-quality raw milk (whether milk is of high-quality or not can be hard to guarantee; please read more on the topic in *The Small-Scale Dairy*, my blog, or any other reputable source) with a good supply of lactic-acid-producing bacteria is capable of fermenting naturally. However, it is most common to add a tiny dose of "starter culture" to clean, fresh milk (chilled or straight from the goat!) to ensure proper fermentation. The lactic acid bacteria break down the milk sugar (lactose) and produce lactic acid. Over 12-24 hours of incubation at moderate room temperature (65-72 F, 18-22 C) enough acid is produced to create a gel. Almost always, a small amount of rennet (milk coagulant) is added at the same time as the culture, helping create a slightly thicker, sturdier curd. The resulting gel is drained, a bit of salt is stirred in, and the cheese is cooled and then enjoyed. Seasonings, savory or sweet, make wonderful additions whether mixed in or sprinkled on. These types of cheeses have a longer refrigerated shelf life, even though they are high in moisture, due to the higher acid level and the presence of beneficial bacteria that make the cheeses less hospitable for spoilage microbes. As with the first technology, the products can be frozen, but cannot be aged successfully. The whey from this process has some nutritional value and can be used wherever a tart liquid is desired (like soups, breadmaking, etc.), or to water acid-loving plants. Chickens will use some, but it's not as appealing to them as whey from the next type of cheesemaking.

Cultured Quick-Set Sweet Curd

This is the longest and most complex approach to cheesemaking and the only one that produces a low-moisture, high-microbe product capable of aging for an extended time. Most of the cheeses we consume fall into this category. Starter cultures are added to warm milk, rennet is added, and within a short time, the milk gels. The resulting curd is cut into pieces, allowing whey to be expelled, and the cut curd is usually gently stirred and often lightly heated to increase moisture loss. Once the right texture and pH (measurable acid level) is reached, the curd is drained and then often pressed. Salt is added either after pressing or before, depending on the cheese style. I won't be giving you a recipe here, but I recommend you try a feta recipe for your first cheese in this category. It's as simple as it gets.

Yogurt, Buttermilk, and Kefir

Unlike cheeses, these products are not typically drained. Other than that, and the lack of any rennet, in essence the process is quite similar to cultured long-set lactic-acid cheeses. Buttermilk is made using starter cultures and incubated at room temperature. Yogurt is made using starter cultures that grow best at higher temperatures (about 110-122F [43-50C]). Kefir—genuine kefir—utilizes kefir "grains" to provide a widely varying spectrum of bacteria and yeasts that culture the milk. Commercially produced kefir is rarely, if ever, made using the traditional method. Instead, a combination of cultures is added. Milk kefir "grains" are a SCOBY, a symbiotic community of bacteria and yeasts, similar to kombucha and vinegar SCOBYs. They must be acquired from a source growing them. They vary greatly in flavor profile and suffer from poor care. There's a lot in my book, *Homemade Yogurt and Kefir*, about the care of grains.

Recipes for Success

I've included a couple of recipes to get you started on the right path. These are my go-to favorites for practical, delicious dairy products. I'm including the use of store-bought cow's milk as an option, in case you don't have goat's milk available yet or don't want to waste it while you practice your cheesemaking skills.

During cheesemaking, all equipment should be extremely clean. Cheesecloths should be cleaned and sanitized just before use by pouring boiling water over them. Milk should be fresh, no more than a few days old. It can even be used warm from the goat! Keep a record of your attempts, that way if you have any issues, or make a particularly awesome batch, you can better understand what happened.

Paneer, a simple high-heat, added-acid cheese, is extremely versatile. It won't melt, so it can be fried, grilled, and used in soups and stews.

PANEER
A high-heat, added-acid curd

If you are at all familiar with Indian food—the curry and chutney kind—then you might have encountered paneer. Paneer is, in my opinion, one of the most perfect cheeses. It's fast, it's versatile (since it won't melt), it can be frozen, and it's higher in protein than almost any other cheese.

Ingredients and Equipment

1 gal (4 l) milk

1/2 -2/3 cup (118-158 ml) apple-cider, white vinegar, or fresh or bottled lemon juice

1/4 tsp (1.5 gm) salt

Heavy-bottomed stainless-steel pot

Part II Meat

Stainless steel spatula

Thermometer

Cheesecloth (ideally with a thread count of 120 per inch; if you can't find the good stuff, double layer looser weave material)

Colander

Steps

1. Pour the milk into a heavy-bottomed stainless-steel pot and place directly on stove.

2. Heat, stirring constantly with a spatula to a gentle boil over 30 minutes. If it takes a little more or less time, that is just fine. (Be sure that you don't overfill the pot, as when milk gets hot it tends to foam and could easily boil over.)

3. Remove from the heat and slowly drizzle the vinegar or lemon juice into the hot mixture while stirring gently—if you stir too quickly, you might break the forming clumps into little pieces.

4. The curds will begin to separate immediately. Keep adding acid and stirring slowly until the liquid portion is a translucent yellow. Adding extra acid, once the whey is translucent, won't hurt, but it won't help either.

5. Let the pot sit uncovered for 10-20 minutes. This gives the curd time to gather and collect, mostly at the top of the pot. It also lets it cool a bit.

6. Dampen the cheesecloth and line the colander. Dampening the cloth helps keep it in place, prevents initial sticking, and allows the cheese to start draining a bit more quickly. Place the colander in the sink or over another pot. Make sure the hot whey won't splash out and burn you!

7. Ladle the curds from the pot into the colander and let drain for 10 minutes. You can pour the whole thing in, but be extra careful about splashes of hot whey.

The Ideal Meat Goat

8. When most of the liquid has drained off, after about 10 minutes or so, gather the curd up in the cloth and squeeze gently to check for any extra whey. Stir in salt. By letting it drain first, you keep the salt in the curd, instead of it being washed out by the whey.

9. Gather three corners of the cloth tightly together and as close to the curd ball as possible. Hold the three corners in one hand. With the other hand, take the fourth corner and wrap it snuggly around the other three, as close to the curd as possible. Each wrap of the fourth corner should be below the previous wrap. This creates a self-tightening knot called a *Stilton knot* (yes, it is named after a step in making Stilton cheese).

10. Place the curd bundle on an inverted plate set inside of a large bowl or in the sink. Place another inverted plate on top of the bundle and then set a heavy skillet or other weight on top of the bundle. Press with about 3-5 pounds of weight for 30-60 minutes. This isn't a very high-tech press, but it works! If you have a fancy press, you can use that if you want.

11. Remove the curd bundle from the press and unwrap the cloth. Your first pressed cheese! You can use the paneer right away, but if you let it chill overnight it will be easier to slice. Keep it in the refrigerator for up to a week or freeze it for use later.

Part II Meat

Chèvre, a cultured long-set cheese, can be drained to several consistencies. Here, well-drained curd has been formed into logs and rolled in seasonings.

CHEVRE

A cultured long-set lactic-acid curd—with rennet

Soft, fresh goat's milk cheese is called *chèvre* (French for "goat") in the United States. It's a simple cheese, and oh so delicious! Sadly, there are many overprocessed versions for sale, particularly imports or domestic versions that are not made as well and have a musky aroma and flavor. While some goat breeds have milk that might be more inclined toward these traits, most goat's milk cheeses should taste clean and fresh. The process for making chèvre (chev-ruh) is identical to that of making the soft fresh cow's milk cheese called *fromage blanc*.

Ingredients and Equipment

1 gal (4 l) milk

1/8 tsp (0.2 g) Flora Danica starter culture

The Ideal Meat Goat

(optional) 1/8-plus tsp (0.7 ml) calcium chloride diluted in 1/8 cup (30 ml) cool, nonchlorinated water; add if using milk that is more than a day or so old or if previous batches don't thicken well

1-2 drops (0.05-0.1 ml) vegetarian double-strength rennet diluted in 1/8 cup cool, nonchlorinated water

1/4-1/2 tsp (1.5–3 g) pure salt

Stainless steel pot or double-boiler

Ladle

Cheesecloth (120 thread count per inch) or other drainable fabric such as flour-sack dish towels or even an old pillowcase

Colander

Steps

1. Warm the milk in a double boiler, or carefully on direct heat, to 86 F (30 C).
2. Sprinkle the culture on top of the milk. Let set for 3 to 5 minutes, then stir gently for 2 to 5 minutes.
3. Stir in the calcium chloride (if using).
4. Add the rennet mixture and stir for one minute.
5. Ripen at 72 F (22 C) for 12 hours or until the curd is just pulling away from the sides of the pot and the top of the curd is covered with about 1/2 inch (1.3 cm) of whey.
6. Cut the mass into 1/2-inch (1.3 cm) vertical columns, then ladle into a draining bag or fine-cheesecloth-lined colander. Tie the corners together or tighten the bag, and hang from the ladle handle or a hook and set across the pot or sink to drain. Room temperature should be about 72 F (22 C) until the desired texture is achieved, usually 4 to 6 hours (longer in a cooler room, shorter in a warmer room).

Part II Meat

7. Empty the mass into a bowl, add salt to taste (start with 1/4 tsp [1.5 gm]), and mix gently but thoroughly, allowing time for the salt to dissolve and disperse evenly.

8. Store it in the refrigerator. You can use it immediately, but the cheese will develop more flavor after a few days, so a period of ripening is helpful. It will keep for at least three weeks.

Yogurt, a cultured long-set curd made without any rennet, varies in thickness depending on the type of milk. It's a great source of probiotics when the right culture is used.

YOGURT

A cultured long-set lactic-acid curd—without rennet

The hardest part about making yogurt is finding a place to keep the incubating milk between 110-120 F (43-49 C), the temperature at which yogurt bacteria thrive. Luckily, there are many options, from high-tech to low. At the low end, you can use a small ice chest lined with a towel and equipped with a couple of jars filled with hot water. At the other end, an Instant Pot or yogurt maker works great too. Incubation times can vary, as can cooling times. It's all very flexible and customizable!

Part II   Meat

Ingredients and Equipment

1/2-gallon (2 l) milk (any kind works, from raw to ultra-pasteurized and skim to cream top)

1/8 tsp (0.5 gm) powdered yogurt cultures or 1/8 cup (30 ml) fresh yogurt with active cultures

Heavy-bottomed stainless-steel pot

Spatula for stirring

Glass jar or jars with lids for incubating

A way to keep it warm (see introduction for ideas; my book *Homemade Yogurt and Kefir* includes an in-depth look at options and the entire process).

Steps

1. Heat the milk to boiling in a heavy-bottomed stainless-steel pot, stirring frequently and watching carefully to make sure it doesn't boil over. Remove it from heat as soon as it comes to a full boil.

2. Cool the milk to 125 F (52 C), stirring to prevent skinning. Setting the pot in a sink of cool water speeds this process to just a few minutes.

3. Add the culture. For powdered culture, sprinkle it on top of the warm milk, let set for one minute, and then stir it in. For fresh yogurt, in a small bowl mix the yogurt with a few spoonfuls of warm milk and whisk until smooth. Add to the remainder of the milk.

4. Cover and incubate at 110-120 F (43-49 C) for 6-8 hours until set. In general, a longer incubation makes a thicker, tarter yogurt.

5. Chill in cold-water bath for 30 minutes then move to fridge to finish chilling. You can even put it directly in the fridge! It will thicken more as it cools, so judge your final results after chilling.

Part II
Meat

A stand mixer with a grinding attachment makes quick work of turning small cuts into ground meat for burger or sausage.

Chapter Six

The Ideal Meat Goat

Meat goats produce valuable products that can be the focus of your goat farm or an added benefit. Goat meat is naturally very lean, bearing some positive benefits, but also some challenges (particularly with mature animals) when processing it into palatable dishes. While all goats can be used for meat, if you want goat meat to be a primary product of your goat operation, it pays to spend some time learning about breed choice, management, marketing, and product harvest.

Kiko bucks enjoying the natural browse at Lookout Point Ranch, Oregon. Photo by Richard Johnson.

Fortunately, for all types of goat breeders, the goat meat market in the United States and Europe continues to expand. This is both because of savvy ethnic populations that are already well aware of the deliciousness of goat meat and expanding domestic palates. Goat meat,

Part II   Meat

also called *cabrito* (Spanish) and *chevon* (A word coined in the US, see sidebar *The Great Mutton Debate*, page 85), has long been a staple in much of the rest of the world, but the Northern Hemisphere has taken longer to embrace this sustainable product. The good news is that the American taste for goat meat has grown so much that the United States imported 70.2 million dollars' worth of goat meat in 2023 (according to tridge.com). These numbers signify a very promising market potential for those considering producing goat meat for the domestic market.

To make a meat goat operation profitable, however, takes planning, experience, and work. Goats aren't trouble-free, totally self-sufficient, hands-off livestock—contrary to popular "wisdom." Growing animals meant to enter the human food stream brings on additional challenges that demand a full knowledge of food safety and the documentation of practices. For an in-depth guide to meat goat production, I highly recommend Oklahoma's Langston University's 2nd edition of the *Meat Goat Production Handbook*, 2015. (Note: At the time of publishing, a 3rd edition of the *Meat Goat Production Handbook* was being readied for publication.)

Goat meat is exceptionally lean. Here it's been sliced thin for jerky.

Goat meat is exceptionally lean, as I mentioned before. This makes it appealing for people looking to lower their cholesterol intake, but it

The Ideal Meat Goat

also makes the meat less tender than a fat-marbled beef steak. In addition, beef is typically hung to age, during which time it grows more tender before it is cut and wrapped, while goat rarely receives the same treatment. For both of these reasons, most goat meat benefits from different cooking techniques that will help make it more tender and pleasing to diners. Regarding flavor, when well-nourished animals are harvested properly, meaning they are not under stress and cleanly harvested—no contamination of gut or bladder contents—then the meat should not be gamey. Most people who've eaten superior quality goat meat liken it to bison in flavor and texture.

Some of the same characteristics that define a good dairy goat (described in Chapter 1) also apply to meat goats, especially those characteristics of general appearance—how the goat's structure is put together so that it can function well. Where it differs most is in regard to the style of bone growth that can be felt on the living animal and seen in the carcass; the type of muscle presented on the animal; and the importance placed on the mammary system. Additionally, an emphasis on hardiness to weather and parasites, fertility, ease of kidding, and mothering skills is likely to be even more important in the case of the meat goat, as it is more likely to be managed with less oversight by humans. Let's go over these differences.

The desirable bone pattern in dairy goats is described as "flat boned." In the meat animal, bones are rounder and thicker. When feeling the legs and ribs on a good meat goat compared to a good dairy goat, there will be an obvious difference. This rounder bone pattern correlates with shorter, thicker muscling. The same is true when you compare a dairy cow to a beef cow. Evaluating the bone structure of the animal will help you decide if it has the potential to develop good muscling. If the animal is already well muscled, you can easily see the results, but if it is young or not in peak condition, its bone pattern will still be something you can evaluate.

Muscling in the meat goat is obviously incredibly important—that's the, um, meat after all! It can be difficult at first to distinguish a well-muscled goat from an overweight goat. Muscling should be evaluated by

Part II Meat

looking at the rear legs, shoulders, and loin (the muscling along the top of the back running in two parallel strips alongside the spine). Fat deposits can be felt behind the elbows and on the sternum, helping you determine if the bulkiness you are seeing is fat or muscle. Feel all of these areas when you are inspecting a goat for its meat potential. In the following section you'll learn how this correlates with the quality of the carcass.

A beautiful meat goat isn't necessarily a *good* meat goat. In fact, the emphasis on the look of the meat goat has led to a loss of hardiness in both Boer goats and breeds such as the Spanish goat when crossed with Boers. Although the Boer goat in South Africa is quite hardy, stock that were initially imported to Canada and the United States sold for extremely high prices and were, according to Linda Coffey in the *Meat Goat Handbook*, "pampered…and as a consequence not as hardy as those raised in South Africa." When looking for breeds to start your meat goat herd, try to choose a breeder who puts their herd to the test and selects for parasite and disease resistance, ease of kidding, and fertility, as well as superior general appearance, bone pattern, and muscling.

Kinder goats, a multipurpose breed originating from Nubian and Pygmy goat genetics. This proud doe mother of triplets is "Murphy" Photo by Haley Schwartz, Maran Atha Homestead, North Carolina.

Breeder Profile Meat Goats
Richard Johnson and Mia Nelson
Lookout Point Ranch, Lowell, Oregon.

Richard Johnson, co-owner of Lookout Point Ranch, feeding two of several hardworking livestock guardian dogs at the Kiko farm.

I first visited Lookout Point in 2016 while researching for *Holistic Goat Care* (see pages 26, 27 for their original profile). I stopped by again in January of 2025 to catch up with where the farm is today. Mia has shifted to working on the couple's adjoining cannabis operation, while Richard actively tends the Kiko goats. Richard, who met me just as a snowstorm was heading into the region, is rightfully proud of the hardy, healthy herd that the 500-acre (202 hectare) ranch now supports.

The Kiko breed originated in New Zealand, developed purposefully to be well suited for *extensive management*. This approach is very hands-off, utilizing natural selection and purposeful culling to breed for disease and parasite resistance and resilience, ease of kidding, superior mothering, strong hooves, and more. This doesn't mean the animals are ignored or uncared for, but it means minimal -handling is the ideal. You won't find any spoiled goats at Lookout Point! Still, health care is not neglected, but is given and assessed. Animals requiring more intervention than others, such as for chronic hoof problems, difficulty kidding, or lack of vigorous growth (under the same conditions as peers who are thriving), are considered for culling.

Lookout Point goats are rotated through large paddocks covering the mountainside property overlooking Lookout Point Lake and Dexter reservoir east of Eugene, Oregon. Livestock guardian dogs play a key role in protecting each paddock from predators roving the adjoining wildlands. Providing hardy, registered breeding stock remains the farm's focus, but plenty of their goats still fill a niche meat market, with some customers using the ranch's ample on-site space to do their own slaughter, whether Western or .

The herd size peaked a few years ago at 125-130 does. This year, however, the couple have bred only ten and are gradually downsizing to a number more comfortable for their current needs. It's good to know that despite their objective, extensive management, Richard confesses that a handful of goats will be sticking around as pets. So, I guess perhaps there *are* a few spoiled goats at Lookout Point!

To learn more about Lookout Point Ranch and Kiko goats, visit https://www.lookoutpointranch.com

Chapter Seven

Meat Goat Breeds

When compared to dairy goats, there are fewer breeds developed specifically for meat production, and there are fewer producers raising these specialized breeds. But as the demand for goat meat continues to rise, so will opportunities for naturally heavily muscled goat breeds. But other breeds can easily provide high-quality goat meat as well. In the profile on page 62 you can learn of a dairy goat farm utilizing their annual crop of male goats to provide value-added meat to the farm's products. Given the prolific nature of goats, and the limited likelihood of finding pet homes for these boys, it's a viable opportunity—particularly where close access to a USDA slaughter plant exists. In these cases, of course, breed doesn't matter. Let's go over the main breeds currently dominating the meat goat scene in North America and much of Europe.

A well-muscled young, colored Boer buck at Sospiro Goat Ranch, North Carolina.

Part II   Meat

Boer

Origin: This very popular meat goat breed was developed in South Africa using indigenous breeds, as well as Angora, European, and Indian goats. From there it spread to Australia and New Zealand, with the first imports to the United States occurring in 1993. Popular in the US as full-bloods (100% Boer ancestry), purebreds (at least one ancestor wasn't a full-blood, but if a doe, it must be at least 15/16s pure, and if a buck, it must be at least 30/32), and percentages (often crossed with Nubian bloodlines).

Size: Bucks 240-300 pounds (110-135 kg), does 200 -225 pounds (90-100 kg).

Distinguishing Characteristics: Floppy ears, arched (Roman) nose, white body with red head and white blaze. Boer goats that are all red or black, called *colored*, are becoming more popular in the US.

Spanish

Origin: A landrace breed (meaning developed over time in a specific area) in the Southeastern United States, especially Texas, that developed naturally over time from original Spanish goat stock brought to the US in the 1500s. Spanish goats were also known as "brush goats." Due to natural selection, Spanish goats are quite hardy, and their parasite resistance is high. The popularity of breeding meat goats, however, has introduced foreign genetics, especially those of the Boer goat, into the Spanish breed. This has led to concern over its survival as a true, or even pure, breed. (When people ask me which meat goat breed I would choose, this is the one.)

Size: According to the Livestock Conservancy there is a wide range of sizes from 50-200 pounds (23-91 kg).

Distinguishing Characteristics: Many colors. Ears are long, but held horizontally next to the head. Usually also a good source of cashmere fiber.

Meat Goat Breeds

Kiko

Origin: Developed in New Zealand with a focus on parasite resistance and excellent maternal traits (specifically for extensive management farming).

Size: Bucks 250-350 pounds (113-159 kg) and does 100-150 pounds (45-68 kg).

Distinguishing Characteristics: All colors. (See pictures in previous chapter.)

Other Breeds

Kinder goats: Mentioned earlier under dairy goats; a midsized multipurpose breed for dairy and meat. (See picture page 56)

Cashmere goats: Listed in the section on fiber goats later on; are often incorporated into a meat goat program thanks to their background of hardy breeds such as the Spanish goat. **Australian brush/Feral/Rangeland goats:** A major source of imported goat meat throughout the world.

Tennessee fainting goats, or myotonic goats: Used for meat and for crossbreeding programs, including the one that created another breed, the **Tennessee Meat Goat™**.

Savanna goats: Developed in South Africa and brought to the United States about the same time as the Boer goat.

Genemaster™ and **Boki goats**: Breeds developed from crosses of Boers and Kiko goats.

Breeder Profile, Meat Goats
Sherwin Ferguson
Mountain Lodge Farm, Eatonville, Washington

Sherwin Ferguson, farm manager Brette Langham, and The Horny Boys

I first met Sherwin in 2010 when she and her daughter Sara came to our farm to purchase their first goats. Like Lookout Point Ranch earlier, this isn't the first profile I've done of Mountain Lodge Farm and their farmstead artisan goat cheese business. I wrote about their cheese, and cheesemaker at that time in *Mastering Artisan Cheesemaking*. Since then, among the many other awards and accolades the farm has received, their aged cheese, called Wonderland, made with the milk of their Lamancha and Nigerian Dwarf herd, took first place at the American Cheese Society competition in 2024.

As with all dairy farms, dealing kindly in an economically sustainable fashion with surplus male goat kids is a challenge. Late in 2023, the farm decided to try raising those males for meat. Already selling the farm's pasture-raised Black Mountain Welsh lamb (processed at a USDA facility) on the farm and at farmers 'mrkets, they knew they would likely have a ready customer base for high-quality goat meat. In the spring of 2024, each male not destined for a job as a breeding buck or pet was designated, get ready for it, a "horny boy." This double entendre applies to these males because the farm leaves them both horned and uncastrated. As Sherwin says, "We didn't want to put them through more procedures than necessary."

The horny boys are raised in an open, tree-shaded pasture with ample shelter and fed both hay and a small portion of Payback Champion Meat Goat pellets. Two weeks before slaughter, they are also fed a measure of corn, oats, and barley (COB). As with the rest of the herd, the horny boys have access to clean water and a free-choice mineral buffet (see *Holistic Goat Care* pages 79-80 for more on this approach that I still champion). During my visit, the boys were all friendly and content, and a few—despite the farm's best intentions—even had names.

The first harvest occurred in the fall when the boys were about seven months old. It averaged a 57% boneless yield of meat per carcass hanging weight. Not bad for dairy goats with part Nigerian Dwarf genetics! The meat, packaged into one-pound packs of ground goat, stew meat, and some highly valued organ meat has a ready market in their Seattle area, bringing $16-17.00 per pound. And best of all, thanks to these kindly raised meat goats, Mountain Lodge is helping open people's minds and stomachs to the deliciousness of goat meat.

To learn more about Mountain Lodge Farm, visit
https://www.mountainlodgefarm.com

Part II   Meat

Chapter Eight

Harvesting Meat

There is a wonderful saying: "Every animal deserves a good life, a good death, a good butcher, and a good chef." This was told to me by a dairy sheep farmer and cheesemaker who loved her animals—both in the pasture and on the plate. The saying really sums up what you must think about both for the sake of the animal and the quality of your experience.

There's nothing like a juicy goat burger to win converts to the small-farm sustainable harvest of your own goat meat. (Meat from Mtn. Lodge)

How and where goats are slaughtered will depend on whether they are for home use, privately sold before they are slaughtered, or for the restaurant and retail market. Federal and regional regulations are very specific regarding the last two of these scenarios and must be

Harvesting Meat

investigated before you start. Perhaps obviously, the least amount of regulation applies to animals raised and used by only the producer, while extensive regulations surround the production and sale of meat to the public. No matter what the intended use of the meat is, you will need to decide on things such as method of slaughter (killing), understanding butchering (the breaking down of the carcass into cuts of meat), proper packaging and storage, and ideas for preparing goat meat to optimize its qualities. I am a fan of Adam Danforth's comprehensive and well-illustrated book *Butchering: Poultry, Rabbit, Lamb, Goat, Pork* for learning the finer details of harvesting goat meat.

Evaluating Goats for Slaughter

In Chapter 6 of *Holistic Goat Care*, I talk about *body scoring* and how it relates to judging a goat's fitness—too thin, too fat, or just right. (See pg 130 of HGC for illustrations.) Goats ready for harvest are evaluated a bit differently than just body condition. Body condition scoring for live animals has a range of 1-5, with the middle being closest to ideal. The numbering system for evaluating goats for meat is called *selection classification* and includes numbers 1, 2, and 3, with 3 being of the lowest quality. The three selections are further broken down into three possibilities of quality from lowest to highest. For example, 3.0, 3.5, and 3.99. It's a bit confusing since the selections indicate that the higher number is the carcass or animal with the lowest quality, but the higher number in that group, for example 3.99, indicates the highest score in that selection.

Other traits that are evaluated on the carcass are the amount of fat covering the animal (subcutaneous fat), the amount of internal fat, and the color of the lean muscle of the goat's flank.

Slaughter and Carcass Terms

Cooler Shrinkage: Loss of carcass weight during chilling. The average is 3-10% from initial hanging weight.

Part II Meat

Dressing Percentage (DP): Comparison of the live weight to the hanging weight. Average DP is 42-52%.

Hanging Weight: The weight after skinning and removal of internal organs (offal), head, and lower leg from hock down.

Live Weight: The weight of the living animal. This number is influenced by not just body condition, but how much it's recently eaten and drank.

Meat-to-Bone Ratio: Comparison of the amount of muscle to the amount of bone on the carcass.

Primal Cuts: Large cuts of meat, often with bone, separated from the carcass. Examples are shoulder, shank, and loin.

Shrinkage: Loss of live weight from the time the animal leaves the farm to when it is slaughtered, for example from dehydration and stress.

Tools for the Meat Harvest

Buckets: To hold blood and offal (internal organs). The quality of buckets depends on if you are saving any of this for processing into products. If so, use food-grade containers that are clean, sanitized, and in good condition.

Butcher knife set: Should include a cleaver or small hatchet, butcher knife, boning knife, breaking knife, shears, bone saw (needed if doing a more complex butcher than what I teach later), and sharpening stone or honing rod.

Captive bolt pistol: Or other properly sized firearm (I use a .22 pistol that uses long rifle rounds).

Cut resistant gloves: Not necessary, but you'll be glad you have them, particularly at the beginning of your learning curve.

Hanging method or work surface: For skinning. Animals can also be skinned on the ground using the hide to carefully roll the partly skinned animal onto its other side (to keep the meat clean) or a tarp.

Hot and cold running water: A hose with a sprayer is handy at the butcher site if done outside or on a floor with a drain.

Nitrile gloves: Or other waterproof gloves.

Working surface: For breaking down the carcass into cuts of meat. A solid, easily cleaned and sanitized surface is important for this stage.

Wrapping supplies: freezer paper, tape, plastic wrap, marking pen. Vacuum-sealing protects the meat from freezer burn (moisture loss while frozen) even better than paper and plastic wrap.

Slaughter

If you've never slaughtered farm animals for harvest, it's likely to be a stressful time in which you grow comfortable with the process and your skills. Just know that with patience and practice it gets better! For your sake and the animals', take your time and ideally have a more skilled person there to help you along. I've taught plenty of people how to do this and always treat it as a sacred moment. I think it's an important experience for anyone who eats meat to participate in, even if it's difficult. Remember, I'm saying this as a thirty-plus-year vegetarian!

When animals experience stress before and/or during slaughter the meat is directly affected. Long, slow stress from such things as being hauled to a slaughter plant, being driven through poorly designed chutes, being prodded and moved by rough handlers, noise, and other things that cause fear causes a chemical change to the muscles of the living animal. This leads to what is known as *dry, firm meat.* Such meat spoils more quickly. Carcasses that exhibit these results are described as *dry, dark, and firm* (DDF). Another negative result occurs when stress is sudden, causing adrenaline to flood the animal's system. If death occurs within 30 minutes of such an experience, the muscle tissue is affected in such a way that it is described as *pale, soft, and exudate* (meaning leaking fluid), abbreviated as PSE. For these reasons as well as for purely humane purposes, the slaughter and all things leading up to it must be as calm as possible.

On-farm kills offer the best possibility for attaining the above goals. The home harvester can perform the kill themselves or hire a mobile service. In some areas government-regulated and government-inspected mobile services exist that can offer the same types of benefits to the animal as well as make it possible to sell the meat to restaurants and at the retail level. In most cases though, the commercial producer (meaning anyone selling meat to the public) in the United States will have to haul the animals to a USDA-inspected plant. I advise you to visit prospective plants and watch their procedures and how they handle the animals before choosing one. Fortunately, thanks in large part to people like Dr. Temple Grandin, who has championed animal welfare innovations at slaughter plants, things have improved. (See the sidebar for more on Dr. Grandin).

Temple Grandin on Ritual or Slaughter

The work of Dr. Temple Grandin through her books, speaking, and website has done much to improve the slaughter methods at commercial plants. I recommend turning to her wisdom and experience when you want to better understand how to improve these conditions for animals. In her 1994 paper *Euthanasia and slaughter of livestock* (Journal of American Veterinary Medical Association, 1994 Vol 204, pgs 1354-1360) she shares her observation of animals being slaughtered by exsanguination, cutting the carotid arteries and jugular veins in the neck (as is required for ritual and halal slaughter), without prior stunning. Here is part of her report (not for the faint of heart to read, by the way).

    Observations of hundreds of cattle and calves during kosher slaughter indicated that there was a slight quiver when the knife first contacted the throat. Invasion of the cattle's flight zone by touching its head caused a →

Harvesting Meat

> ← bigger reaction. In another informal experiment, mature bulls and Holstein cows were gently restrained in a head holder with no body restraint. All of them stood still during the cut and did not appear to feel it. Disturbing the edges of the incision or bumping it against the equipment, however, is likely to cause pain. Observations by the author also indicated that the head must be restrained in such a manner that the incision does not close back over the knife. Cattle and sheep struggle violently if the edges of the incision touch during the cut."
>
> As an addendum, I've recently been told, and have read, that some halal butchers are using a captive bolt to stun the animal first. The stunning method, whether bolt or other, must be reversible in order to comply with halal rules, but this is encouraging news for animal welfare activists and even the butchers doing it. We must keep in mind that when things are calmer and more human for all food animals, the situation is likely to be kinder to the workers as well.

In commercial plants that are not ritual slaughter facilities, animals are restrained and then a penetrating captive bolt pistol (which uses gunpowder but no bullets) is used to stun or completely kill the animal. After confirming the animal is unconscious or dead, its throat is cut and the body positioned so that the blood will be effectively drained from the carcass. The heart muscle will continue to beat for a short period of time after death which helps drain the blood. A .22 caliber or larger (such as .38 or 9mm) firearm can accomplish the task as well—but is more likely to result in instant death, though, than stunning. But using a firearm includes an increased risk for workers and the need for further training and perhaps licensing. A firearm and/or a captive bolt can be purchased for about the same cost, and rounds (blank in the case of the captive bolt) are fairly inexpensive. The captive bolt gun, however, is not cheap, and neither is a firearm. If you know other farmers locally, there's the possibility of pooling resources to purchase one.

The home processor and mobile farm killer is more likely to use a firearm, often because one is already owned to protect livestock or perform on-farm euthanasia. A skilled slaughter professional will likely use a rifle in order to be able to dispatch an animal from a distance, such as a steer not accustomed to being handled. This removes the trauma of capture for the animal.

On the farm, once the animal has been bled out, it can be hung for the next step. This is done by exposing the strong tendons just above the animal's hocks (the Achilles tendon, sometimes called the *gambrel cords*) and then inserting hooks, or other means of suspending the animal, through the opening. Many things can be used to lift the carcass: a tractor with a hydraulic bucket, hay hooks strung on twine and suspended from a rafter, an engine hoist, etc. If you aren't able to hang the animal, it can be processed lying on sloped ground, with the head lower than the body, in a clean area, on a tarp, or placed on a table.

Butchering: Stage One

Butchering begins with the initial stage of gutting and skinning. These two steps can be done in either order that makes the following stages easier for you. In the instructions below, I skin first, then gut. Hunters will be familiar with field-dressing, where the animal is gutted where it fell and then transported to another site for skinning and the second stage of butchering.

Before we start on this involved process, I want to thank the young kid that gave its life for the lesson and the meal that followed most of these photographs. It's never easy to take a life, (and it shouldn't be!) much less that of a young animal. You must remind yourself that as a

Harvesting Meat

farmer and home processor you have been gifted with the opportunity to provide a good life and gentle end for the animal, especially when compared to a commercial facility. On that note, starting your meat harvest education with a small animal is in most other ways far easier than doing so with a mature goat.

If you are new to this process, it will initially take some time, but your skills will rapidly improve! I suggest not worrying about saving the hide on your first go, or at least not being too disappointed if you don't get a perfect hide removal. Use a sharp boning knife to make the cuts and try to keep your dominant hand clean, using your other hand to pull and hold the hide as you work. You will have already opened up the skin around the hocks to expose the tendons when you hung the goat, so continue your work by skinning the rest of the rear legs. Always cut from the inside outward so as not to drag contaminants onto the carcass. Remember, the inside of the animal's body is perfectly clean and free from contaminants at this point! Make cuts as shallow as possible to not puncture the abdominal cavity (risking stomach, gut, bowel, or bladder content contamination) and to keep as much fat on the carcass as possible (which will help prevent it from drying out during the butcher).

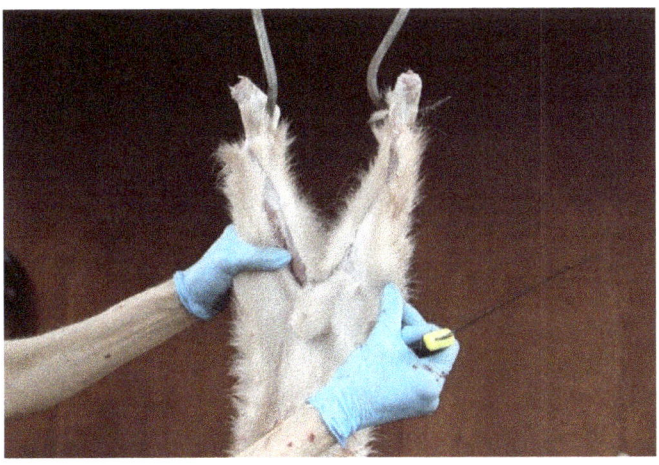

After the lower leg has been removed, the goat is hung using two repurposed hay hooks, then a cut is made on each leg that meets in the middle.

Part II Meat

A note: When an animal is dispatched, it can be completely dead and still have movement. Legs will paddle, muscles will contract and release, eyelids will twitch. This movement is normal due to the sudden loss of connectivity between the brain—the electrical command center—and the nerve impulses running the muscles. It's quite disturbing to see, especially if you are not expecting it. But, in fact, if you've attempted the kill with a shot to the animal's brain, and the movement DOES NOT occur, there's a strong chance the animal is not dead, but only stunned. Quick exsanguination will remedy this. But if in doubt, keep a stethoscope around so you can listen for a heartbeat.

Skinning Step-by-Step

1. Skin the lower leg an inch or so more toward the hoof and then cut the lower leg off below the hock joint. Don't cut too high or you could cut the tendon by which the goat is hanging! (Trust me, you're probably going to do this at least once.) Cutting through the joints is difficult. If the goat is a young kid, you can use large pruning shears or bolt cutters. Or use your bone saw or another type of saw. If not, work your way slowly through the joint, bending it to open space between the bones.

2. Cut through the meat around the head at the highest point on the neck (which is the lowest when you are looking at it hanging), then use your knife to cut the tendons that attach the head to the neck. Move the head around a bit to help the knife work its way through. There's a sweet spot just above the first vertebrae where you can easily sever the connection without ruining your knife. You can remove the head completely at this point or later. I always do it now. It just seems more respectful, if that's possible. (If you want to harvest the brain for brain-tanning, see page 148 for instructions.)

3. Hang the goat from the hock tendons at a good working level for you.

Harvesting Meat

4. On each rear leg make a vertical cut outward through the hide and then connect the two cuts at the groin, in front of the mammary and anus.

5. Holding the blade so that it doesn't penetrate the abdominal wall, cut down the abdomen from the groin to the neck with two long cuts about 2-3 inches apart. Pull and cut this strip from the carcass.

6. Make an incision from the bottom of this last cut down to the neck where you made the incision for bleeding.

7. Run your hand under the hide and lift it away from the animal (called *fisting*). On younger animals, this is very easy compared to older animals.

8. Isolate the tail and cut a small ring around its base. I usually remove the entire tail at this point.

9. Continue removing the hide by holding and pulling with your dirty hand and gently cutting the tissue holding the hide to the body. Use long, smooth strokes. Start at the groin and work your way down the body. This keeps the dirty hide from touching the carcass. When the hide has been opened so 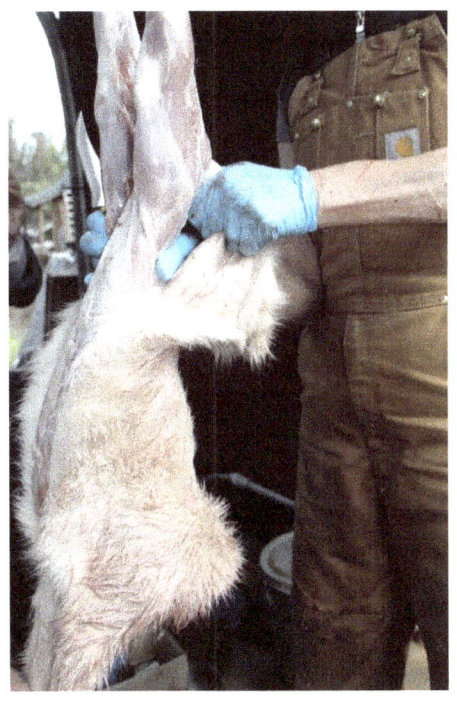 that it exposes both legs and inner thighs you can use your fist to reach between the hide and carcass just in front of the rear legs and around toward the back of the animal, it should separate relatively easy. Then push your hand upward along the animal's back and out toward the cut around the groin.

Part II Meat

10. Skin the front legs down to the knee joints (carpus) then cut through the knee joint at the top and remove the lower leg and hoof. (You can skin down to the pastern and cut there instead.) As a reminder, all joints are more or less difficult to cut through—there's no straight cut that works. You just have to work the knife between articulated bones. (Remember, bending the joint back and forth as you navigate it is very helpful.)

11. Pull or cut the hide off down to the head, then remove the head if you haven't already.

Gutting (Eviscerating) Step-by-Step

1. Cut around the anus, holding it with one hand, then use a piece of butchers 'twine or other string to tie it shut. This helps prevent contamination of the carcass.

2. Now pinch the skin just below the anus and pull it outward, then make your first small cut into the abdomen, being careful to not cut into the bladder which is located just inside.

3. When cutting into the abdomen fully, you will have to be **very careful** to not puncture the digestive tract or contamination will occur. To prevent this, you can hold your hand inside the abdominal wall and hold the knife with its sharp end pointing outside of the animal, rather than toward the innards, and then cut downward. Cut straight down until you reach the sternum.

5. If your process has taken a while, the rumen gasses will have continued to expand and at this point the rumen may want to come tumbling out, so be prepared with your bucket or container. Look for the bladder and rectum inside the opening and pull them all outside of the animal, then cut the bladder free. **It's very important to not let it burst or be cut as the urine, particularly from an adult buck, will taint and flavor the meat.** Remove any organs you plan on saving (with the kid used for the photographs, we saved the abomasum and the stomach and made amazing traditional rennet for cheesemaking), then pull the entire mass out and into the bucket. The

Harvesting Meat

esophagus may still be connected (unless it was severed during exsanguination). Locate it and move your hand downward around it toward the goat's neck. You can use a towel to get a good grip on it if that helps. Pull it up and outward in one quick motion.

6. Cut the tissue of the diaphragm to expose the lungs. Reach into the space and grasp the trachea (as with the esophagus, be sure it has been severed at the neck), then use your other hand to grab the lungs and heart and pull them up and out of the chest cavity. Cut any blood vessels that are still attached. Remember, if the animal was properly bled out, there should be little blood remaining in the vessels. Expect a few clots here and there, but this won't be a bloody mess.

7. At this point if there are any hairs or other contaminants on the carcass, rinse the carcass with cool water and a sprayer. Rinse inside as well. That's it for the initial carcass prep. It is now ready to be aged or butchered into smaller cuts.

Tips for Gutting

You have two options for when to remove the insides of the goat. It's always fine, and probably most common, to do this before skinning. In fact, if you know you are going to take some time for skinning, this will reduce bloating in the carcass. But if you are quick, skinning can also be done first.

Although the animal is dead, the microbes in the rumen will live for hours longer, continuing to produce gas.) For this stage, you will want to have a plan for what organs you are going to harvest for food and what will go to the compost pile, raw dog food diet, or other disposal method.

Part II   Meat

Aging

After death the muscles of all animals stiffen in what is known as rigor mortis (this is different than the initial violent contractions that occur after sudden brain death). Aging the carcass for a few days allows for this phenomenon to pass. When aged even longer, muscle proteins break down into a more tender structure and other flavor compounds develop. Meat aged for just the right amount of time will be more tender and flavorful than fresh meat. Beef, higher in fat and with a nice protective layer of surface fat, is often aged for several weeks and poultry just for a couple of days, while goat should be aged for up to two weeks.

Dry Aging: Dry aging is usually done by hanging the carcass in a cold space (40 F/4.4 C) where the relative humidity (measured by a humidity gauge) is between 70 and 80 percent. In large meat lockers where beef carcasses are hung, fans are used to keep the air moving and help dry the surface of the carcass to keep it from being too wet. The meat will lose some moisture. During aging, harmless white molds will grow on the surface and the carcass will darken. The mold and hardened surface must be trimmed and scraped clean before packaging. Because goat meat is so much leaner than beef, it is prone to becoming overly dry if aged this way. To help prevent this, rub some of the excess internal fat to over the carcass or wrap it in oiled cheesecloth to help protect it from over-drying. Dry aging at home can be done using a refrigerator without a freezer compartment (which provides more room inside the unit as well as lacking the over-drying tendency of a freezer's fan blowing air into the refrigerator section or, if the weather is cold, in any clean, cold space that is safe from pests and predators—including house pets.

Wet Aging: Wet aging is easier and more common with goat meat. The carcass is first hung for a few days to allow rigor mortis to cease, then the animal is broken down into larger pieces which are vacuum sealed in plastic and then refrigerated and stored for aging. My hands-down favorite way to age goat meat is a form of wet aging that utilizes a plain brine in which the cuts are soaked. It's the best of both worlds, allowing salt and time to create a superior texture in the meat.

> **Brining Goat Meat**
> 1. Butcher into primal cuts. Rinse well.
> 2. Mix a 1 gal:1 cup mixture of salt to water by volume. For example, 2 gallons of water to 2 cups of salt. Dissolve well.
> 3. If you want to add seasonings, go for it. I like a bit of brown sugar, apple-cider vinegar (use 1/2 the amount of salt used for each), bay leaves, and peppercorns.
> 4. Submerge primal cuts (or smaller) in brine. Be sure no part is exposed. (I use Ziplock bags filled with water to weigh down the top.)
> 5. Put in refrigerator or ice chest for at least four days.
> 6. Drain and use.

Butchering: Stage Two

There is no one right way to break down the goat into smaller cuts (called in the larger meat industry, fabrication). Much depends on your own desire to make beautiful, identifiable cuts, such as chops and roasts; your skill level; the tools you have available; and what you intend to use the meat for. If you are making mostly ground meat or sausage, for example, pretty goat chops and a rolled boneless shoulder roast are irrelevant to the butcher. Fine butchering is a skill and an art. Fortunately, if it's your desire to master this, there are several beautiful books out on the topic including the earlier mentioned *Butchering* by Adam Danforth and Meredith Leigh's *The Ethical Meat Handbook* (where you can reference the lamb section as equivalent to working with a goat carcass). Let's go over the simplest way to process the carcass into usable cuts.

To quickly and with minimal skill break a goat carcass down all you need is a flexible, sharp boning knife. The process, sometimes called *muscle boning,* focuses on cutting the meat from the bone and packaging it boneless. You have the option of keeping the muscles from the legs in one chunk and then rolling it into a boneless roast or separating out the individual muscles and cooking them more like steaks or using them for stews. If you don't have a lot of freezer space, this method is especially helpful. We had some good freezer space at our farm, but I still preferred this method. We saved most of the bones, which will still have some meat on them, for raw dog food or cooked them down right away for meat stock. Here are the steps for butchering a goat using the muscle-boning approach.

Fabrication Step-by-Step

1. Lay the carcass on its side on a clean work surface.
2. Remove the front legs.
    a. Cut under the front leg from the chest up along where the neck joins the shoulder to free the connection. The shoulder is only attached by tissue. There isn't a bone joint.
    b. Cut the tissue connecting the shoulder at the withers.
3. Bone out the front legs.
    a. Cut the meat from the shoulder blade and leg bone by making a cut through the muscle and along the bone to open the muscle and expose the bone.

Harvesting Meat

      b. Cut the exposed bone from the meat.

4. Remove the back legs.

    a. Move the leg a bit and locate the hip joint. It's a ball and socket joint just like that of our own hip, but with a surprisingly small femur head (the ball part). Use your knife to cut around the muscle and tissue that connects the joint.

    b. When the hip joint is free you can separate the leg from the pelvis.

5. Bone out the back legs.

    a. On the inside of the leg, run your knife between the two large muscles to expose the femur.

    b. Continue to separate the muscle from the leg below the stifle joint and around the front of the femur to remove the bone.

6. Remove the loin (also called the *backstrap*) along each side of the spine on the outside of the animal:

    a. You can feel the loin along the animal's back. Slide the knife along the spine between the bones and the muscle and cut down the spine.

    b. Hold the loin muscle up and free it by cutting through where it rests against the ribs.

7. Remove the tenderloin located alongside the spine similar to the loin, but on the inside of the body cavity.

8. Trim other meat from the neck, flanks, and wherever you can for stew or ground meat.

9. If you want to break the carcass down further for dog or soup bones or disposal, a bone saw or even an ax can be used.

Part II   Meat

Packaging and Storing for Home Use

Freezer burn, the deterioration of the meet when exposed to freezing, causes oxidation and drying. To limit this, you should double wrap or vacuum seal the meat. I use plastic wrap to cover the cuts, then wrap that in freezer paper. Use freezer tape (it looks like masking tape but will stay in place in the cold of the freezer) to tape the packages shut. Then label each package with the date and cut of the meat.

Place the packages on a rack or tray in single layers and place in the deep freezer (aka, chest freezer). This ensures that all cuts cool and freeze evenly. Although food preservation experts and government guidelines say that properly frozen meat will keep for 9-12 months, most anecdotal evidence indicates that it will last much longer if: the deep freezer is set to its lowest temperature (0 F [-18 C]), there are no temperature fluctuations (from events such as power failures or defrosting), it was wrapped properly, and it is subsequently cooked properly. While this isn't a suggestion or recommendation, in evidence, we are still going through beef that was packaged 3-4 years back and the quality is excellent (according to the meat eaters here at the farm, that is!).

Nubian goats, while primarily used as milk goats, are known for their sturdy muscular build. This makes them especially good for meat and crossing with other meat breeds.

Harvesting Meat

> **My First Goat Harvest**
>
> The first goat I butchered for our family was an adult doe that was obnoxiously noisy, not a good milker, and not a good show goat (in other words, there were not good prospects for finding her a quality new home). The family was feeling less than enthusiastic about eating her—until I turned the first cuts into jerky and handed a slice to our youngest daughter who was about fourteen at the time. She took a tiny nibble, then another. Her eyes widened, and she asked, in her usual rather nonpolitically correct fashion, "Is there another one we can off?" Translation: it was delicious. Since that time, I have used our goat meat for sausage, stew meat, burger, roasts, and grilled cuts. People tell me it tastes like mild beef or bison. It has no gamey flavor and when prepared properly, is not at all tough. Even an adult buck who was harvested was deemed, by my picky kin, to be mild and delicious.

Part II   Meat

Chapter Nine

Marketing Goat Meat

Some producers are able to take advantage of growing or robust markets for their well-raised goat meat, such as Mountain Lodge Farm, profiled in Chapter 7. In this chapter, we'll go over some of the things that might help you in this endeavor. Please know that laws vary state to state regarding the legality of selling meat directly from the farm, but in all cases, if meat is to be sold across state lines the USDA rules will override any state or local exemptions.

USDA-inspected goat meat being sold at a farmers' market. Photo by and courtesy of Mountain Lodge Farm.

Marketing Goat Meat

Food Safety and Quality Assurance Programs

In Part I on milk, I deferred the discussion of setting up your production to legally sell milk and/or cheese to my other books. Since the opportunity to sell meat, or animals for meat, is even more accessible (and I haven't written a book on it), I wanted to give you a glimpse of some of the considerations. Don't forget, I highly recommend Langston University's *Meat Goat Handbook* for a more in-depth discussion on everything to do with marketing goat meat.

The level of documentation that you will need to do will depend on the regulations imposed on your facility. Those will likely be dictated by the size and scale of your operation, including how widely your products are distributed, and the laws of the local and regional jurisdictions. The home meat producer must also have food safety on their mind, but the documentation requirements are completely up to you. Food safety programs initially sound complicated and daunting. Like any new skill, though, once you learn the terminology and start understanding the principles, things start making sense.

The first two terms you will encounter when talking about food safety programs are best management practices (BMPs) and hazard analysis critical control points (HACCP—said "ha-sip"). First, rest assured that best management practices are really what the entirety of *Holistic Goat Care* and other thorough goat health guides are about. They deal with everything to do with raising healthy, well-cared-for animals that, when they enter the food supply, become healthy, safe food for humans—and pets too. HACCP is an official program whose goal is the reduction of risk and the documentation of how that is accomplished. Unless your meat production program is large-scale, you probably won't have an official HACCP plan (or other audited-type program), but the principles of these plans should be applied to any size farm.

Other organizations such as the American Sheep Industry Association and the beef industry have official programs that give a standardized model for quality assurance. A similar program, called the

Part II   Meat

Quality Meat Goat Producer Certification, is listed through Langston. Their *Meat Goat Handbook* (see information in back) is complementary to the information on the website (which can be viewed for free if you're not taking the certification test).

Markets and Marketing

There is far more goat meat imported than is produced here in the United States. The number of goats processed in the US in 2014 compared to 1991 has only a little more than doubled, while the amount of goat meat imported has gone up almost 20 times what it was. If the current demand in the US were to be met by domestic producers alone it would mean tripling production. Most of the goat meat imported, however, is from rangeland, aka feral, goats in Australia. The quality of the meat harvested in such a manner is likely to be far less (due to all of those issues surrounding stress at harvest that I talked about earlier) than that of the small producer able to ensure a less traumatic end for the animals.

  The market for goat meat is likely to continue to grow. And hopefully, consumers will realize that not all goat meat is equal! As awareness grows, so should a market for high-end goat meat. Interest in local foods and small-scale farms along with the attention that goat meat is getting from many top chefs as well as trendy food trucks and food writers should nurture this trend. Let's go over the different ways to get your goats to the market and who your customers might be. A terrific website that allows you to search for different types of markets is Cornell University's sheep and goat marketing website. It isn't a complete listing by any means, but it might help you find a market in your area. At the end of this chapter, you'll find a sidebar listing terminologies and target markets for different ages of goats. These listings are quite variable in some regions, but they'll give you an idea of the importance of figuring out what the market is and the need to determine what "type" of goat you can most easily sell at any given time of the year.

  The USDA has defined what are called the Institutional Meat Purchaser Specifications, or IMPS. If you choose a market that includes you in the decision-making as to what cuts to produce or if you are

Marketing Goat Meat

sending your goats to a processor supplying this market, you will want to review the categories and recommended primal cuts. You can access that information through a publication available online by Louisiana State University's Agricultural Center called *Meat Goat: Selection, Carcass Evaluation, and Fabrication Guide*, on the USDA's website, or find it reproduced in Langston University's *Meat Goat Handbook* (I think you're getting the idea I really like this book...).

Animals Direct to Private Customers

There are several options for selling animals destined for meat directly to buyers. This can be profitable, even if it isn't a prime part of your market. However, it requires more coordination, interaction, and a loss of some of your privacy. Here's a summary of the options.

The most obvious option is selling the animal and then having the buyer take it away. But this is arguably stressful for the goat as it is traumatic for them to be completely removed from their herd. If you sell in groups or pairs, however, it will be less so.

Custom-exempt slaughter: In most, if not all, states in the US, customers can pay for an animal before it is slaughtered. The animal can then be dispatched by a mobile service and taken to a slaughterhouse and butcher, at which point the customer can take it to their own home for processing or have it processed on the farm. Please check into the regulations before making any of these options available to buyers.

On farm harvest: In this case, you allow the buyer to do their own harvest, but on your farm. If you decide to allow slaughtering and butchering on your property, it's a good idea to provide a space for the process. This might be as minimal as a tree branch for hanging and a hose for washing. Insurance coverage for liability will also need to be investigated. If you believe your customers will be from a religion that requires animals to be of a certain age and quality, those are important to consider as well.

Part II   Meat

The Great Mutton Debate

In her book *Goats in America*, Tami Parr covers the convoluted history of American's struggle with just what to call goat meat. We have "pork" for pig meat, "beef" for cattle meat, "poultry" for chicken meat, after all, so what about goats? It was in 1922, Parr says, that the Texas Sheep and Goat Raisers Association came up with the term *chevon*—a combination of the French words (goat) and *mouton* (sheep). It made sense. Consumers were familiar with the term "mutton" for adult sheep meat (as opposed to lamb) and, indeed, prior to the convention, it was common for butchers to package and label goat meat as mutton. As goat meat became more common, though, consumers were often astounded to find that the mutton roast they purchased, believing it to be sheep, was actually goat meat.

A decades-long kerfuffle ensued, with accusations of deceptive labeling and the consumers' lack of distinguishing palates (in other words, "they won't know the difference"). A bill even made it to the floor of the US House of Representatives requiring goat meat to be called, well, goat meat. But the debate continued clear into the 1970s when the finally required that goat meat be labeled as either goat meat or "chevon. For the full colorful story, please check out 's book, *Goats in America: A Cultural History*.

Marketing Goat Meat

Meat Direct to Customers, Retailers, and Restaurants

Meat that is to be purchased by the public, whether in a package or as a part of a meal, must be slaughtered and butchered in an inspected plant. In the United States the USDA oversees these facilities. In a growing number of states in the US, meat, and even more often poultry, can bypass USDA oversight and be processed in facilities overseen only by the state. See the resources for a link to the USDA site to learn if your state has its own meat and poultry inspection (MPI) program.

Meat processed in USDA facilities can be sold at any level of retail—from farm store to grocery store and restaurant to wholesaler. In the case of MPI processing, the sale can only occur in-state, unless a cooperative interstate shipment program exists, allowing the meat to be shipped across lines to a neighboring state. A farm can decide how cuts are to be made, packaged, etc. Things such as custom sausages and bratwurst are quite a popular way to get goat meat onto the plates of diners new to the product. And thankfully, more chefs are eager to source meat directly from local farms where they have a perceived oversight of the practices and can convey that quality via their menus and marketing.

If you decide this approach is a part of your plan, you will need to understand pricing. Four levels of pricing are possible with variations within each: direct retail price, wholesale price, distributor price, and wholesale to restaurants price. There are formulas for all of these, but you should start with what is already common or acceptable in the market, if that information is possible to find out. You will also likely need to acquire product liability insurance. Some retailers and distributors will require different levels of minimum coverage. Some of them will also require that you have a food quality assurance program in place. In fact, having a written food safety plan might even save you a few dollars on your liability insurance!

Part II   Meat

Animals to Live Animal Market and/or Auction

In this type of scenario, customers select living animals from pens and purchase them. Then they are slaughtered on-site. Typical buyers are ethnic and/or from religions that require particular methods be used and animals selected.

Livestock auctions are not for the faint or tender of heart. Although they used to be common throughout rural America, relatively few are still in operation. Taking animals to auction is the simplest way to sell a lot of animals quickly. That being said, you will have very little say on the price and the way the animals are handled. If you have auction options, spend some time investigating their practices and be pound or by the head, and how, and when, you will be paid.

Animals Direct to Dealer or Meat Packer

Selling your animals to a buyer that then takes on all of the responsibilities and costs is a simple way to move a lot of animals. Of course, your profits are lower, but so are expenses. In this scenario livestock dealers and meat packers come directly to the farm and purchase and pick up animals. Buyers might be large processors or small. In fact, becoming a buyer is something a goat producer can do as a way to maximize their own trip to the slaughter plant and produce more product. (I have friends that do this on a small scale, picking up wethers and bucks from producers they know in order to make the trip to the slaughter plant more cost-effective for everyone.) The dealer or packer may pay by the head or by weight. An on-farm scale for weighing live goats is important to have when being paid by weight.

Marketing Goat Meat

The Lingo of Special Markets for Goat Meat

Newborn Kid or Shoebox Kid: Very young. Often only a week or two old. Usually milk-fed and not yet ruminating. Often given as gifts for religious rituals, hence the term "shoebox." Pygmy goat kids and small breeds have a market opportunity in this category.

Suckling Kid or Easter Kid: Called *capretto* (Italy), *katsikia* (Greece), or *cabrito* (Hispanic market). Dam raised; non-disbudded kids are preferred. Usually 4-12 weeks of age and between 16-40 pounds. Easter, Christmas, Hanukah.*

Cabrito: Milk-raised kid, usually 25-40 pounds.

Market Kid: Weaned kid usually under a year old (for the Muslim market, kids must not have any adult teeth). Ideal weight 60 pounds. Male or female is fine. The Muslim market doesn't prefer kids with excess fat.

Chevon: Any age animal, but most often those under 2 years of age and over 60 pounds. Not for any particular market.

Cull Does and Bucks: Any age, any gender. This stronger flavored, more complex meat is preferred for many ethnic stews and curries.

*Ethnic holidays often occur on changing dates. You can access the current year's calendar of ethnic holidays with an internet search. I recommend this as you develop your market plan for the year.

Part II   Meat

Wholesale Meat to Distributor or Large Retailer

Distributors are companies that take orders from restaurants and retailers. When they purchase your product your compensation is from the lowest tier, but you also have greatly reduced costs and labor related to marketing and customer support, including possibly spending days on the road delivering product. Some large retailers might be willing to act as their own distributor if your goat meat is highly desirable. In that case, you may be able to deliver a large order to one store and they will disperse it to the other stores in the chain.

Timing Goats for the Market

As you can see from the sidebar on special markets, there are a multitude of moving targets for holidays that are great for selling goats for meat, not to mention the many stages at which the goat is considered prime for those markets. Before you decide how to plan your production year, it's a good idea to sit down with a notebook or a spreadsheet and coordinate the target sales date with your breeding and kidding dates. Ideally, you will be able to spread the harvest and income out over the season, unless you plan on selling the crop *en masse* to a single buyer.

(To add to the confusion, the Islamic calendar is not the same as the more common 365-day Gregorian calendar year. The Islamic calendar used by Muslims is about 10-12 days shorter than the Gregorian, so important holidays in which goat meat plays a starring role don't coordinate with the same dates on the more common calendar.)

Breeder Profile, Fiber Goats
Al and Linn Schwider
The Pines Farm, Maple Valley, Washington

Linn in her on-farm fiber store.

For the past four decades, Lin and Al Schwider have invited the public to the Pines Farm for their fall fiber festival. Located just far enough east of Tacoma and Seattle to feel like an escape to rural life, the festival brings spinners, families, and fiber enthusiasts from the surrounding area. It also gives Lin and Al a chance to show off their hard work—and believe me, there's a lot of it. On the day I visited, Al was multitasking a repair to the well and cleaning and bedding one of their barn's big pens. The fifteen-acre farm has been owned by the Schwiders since 1978. It was in the early 1980s that they bought their first Romney sheep, a meat and wool breed which they are still known for. Twenty years later, they added goats after a nearby breeder gave Lin a single kid to raise. Already having a gifted eye for fiber, Lin was drawn to the Angora goat's soft mohair.

The Pines Farm currently includes twenty-three does and three bucks. I arrived just at the start of kidding season, and lucky me, because if I'd thought baby goats couldn't get any cuter, I learned I was wrong! With soft, curly locks fringing their eyes, baby Angoras are irresistible. The

Pines Farm dam-raises (meaning no bottle feeding) each doe's offspring, averaging 1.5 kids per doe (the norm for Angora goats). Lin usually sells most of the kids, but this year she will be retaining most of the females, having just brought in two new bucks (with new genetics) last fall. These new animals originated from a herd in Texas, where most of the Angora goats in the US are raised. The Pines Farm adult does average 4-5 pounds of fleece per shearing (which occurs twice per year). Kids have their first shearing in the fall, with an impressive yield of 2-3 pounds. This first shearing yields the highest value fiber. The Schwiders market these kid fleeces themselves. (Although I don't spin, I couldn't resist purchasing a couple of bags of locklets myself.) Lin, with a degree in art (and a house hung with a gallery of her own paintings), takes great pride in how she processes the fiber, particularly in her custom palette of gorgeous colors. The farm has a lovely on-site store that sells finished skeins, raw fleece, rovings, batts, kits, patterns, and a few finished items. They also ship via their website (see link below).

Shearing is done by Al and Lin's adult son and daughter. Their daughter, Amy, is also a judge, taking her micron meter to shows to help breeders assess the fineness of their herds' mohair. In addition, Amy optimizes the two-year-old and older doe fleeces by cleaning, dyeing, and marketing them for use by crafters as hair for dolls. (Really, she's not the only one. Check it out on Etsy and Amazon.) Lin mentioned one thing that is at odds with the literature on Angora goats (as with most things documented, I tend to believe the anecdotal evidence presented by those working in the field). She said she's seen no correlation between high nutrition and a decrease in fiber quality. That's good news, as, of course, good nutrition leads to increased fertility and health. If it doesn't decrease fiber quality, it's a win-win.

For more on Angora goats, to purchase some of the Pines Farm's products, or to learn more about their fall festival, visit https://thepinesfarm.com.

Chapter Ten

Meat Products Primer

Goat meat is unique from beef in several important ways, most importantly, the lack of marbling and its special needs for aging. Other than that, we've found it can replace beef in most recipes, except for, perhaps, if you just want a juicy oversized steak! Nutritionally, goat meat has almost the same amount of protein as beef, but is lower in fat (and cholesterol).

Me and more than one friend checking progress on a Pholia Farm goat roast, 2017.

Part II Meat

Goat meat is considered "red" meat, but, as I've mentioned a few times, unlike beef, the muscle tissue of goats lacks "marbling," the deposit of intramuscular fat (IMF). It's the deposit of IMF that cattle excel at—and the reason for beef's succulent, flavorful reputation. Even within beef breeds, there are different genetic propensities for developing IMF. For example, Wagyu cattle are famous for their ultra-marbled meat. This is true in goat breeds too, but in general, all goats will have far less marbling in their muscle tissue than most other red-meat species. For a market seeking leaner meat, this is a win! But it might require an adjustment of expectation for cooking and consuming.

In Chapter 8 I mentioned that dry-aging goat carcasses to achieve a more tender meat is rarely done because of the animal's lean nature and small size (meaning less profit for an aging facility giving up space to such a small yield). If you want to try it at home, though, the directions are in Chapter 8. There I also talk about wet aging and my hands-down favorite method of tenderizing goat meat: brine aging.

Cooking and Using

A sampling from chef Stephanie Izard's Chicago and Los Angeles restaurants, Girl and the Goat (such as "confit goat belly with butternut squash, green apple sweet 'n' sour, and pecan crunch") reveals the growing popularity of goat meat, as well as the myriad ways it can be served beyond curry. A quick search for goat meat recipes on the bookmarking website Pinterest will give you immediate access to hundreds of recipes and tips on cooking goat. Some traditional recipes rely on a lot of spices, but high-quality, cleanly processed goat meat can shine with minimal seasoning. If you must work with goat meat that has not been aged or brined or comes from a source (such as imported feral goat meat) where the meat might be tough due to circumstances surrounding the harvest (see page 67 for more), then marinating, brining, and long stewing are all methods for improving texture.

Recipes

Here are a couple of recipes to get you started.

Goat jerky ready for snacking

JERKY

Goat meat makes awesome jerky thanks to its naturally lean nature. It's a great way to turn small pieces into delicious, nutritious road snacks! This recipe is an easy one to modify to suit your tastes.

1-2 pounds (450-900 g) of goat meat

1/2 cup (120 ml) tamari or soy sauce

1/4 cup (60 ml) balsamic vinegar

2 cloves garlic or 1 teaspoon (5.5 gm) dried garlic granules

1 teaspoon (3 gm) onion powder (you can use real onion, but remove after marinating)

1/2 teaspoon (1 gm) ground black pepper

1 tablespoon (21 gm) honey

Part II   Meat

1/2-1/4 teaspoon (1-2 ml) liquid smoke (optional)

1 teaspoon (5 ml) hot sauce such as Cholula (optional) or 1/4 teaspoon ground cayenne pepper

1. Pre-freeze sections of lean, gristle-free meat.
2. Thaw partway and slice into long, thin strips lengthwise of grain. (If you cut it cross grain, it will be more likely to fall apart.)
3. In a bowl, mix the other ingredients until well blended.
4. Add the meat slices, cover, (you can place everything in a sturdy, Ziplock-type bag) and marinate for 2-3 days. Stir one to two times a day (if using a bag, just turn the bag and squish everything around a bit).
5. After the meat is well marinated, drain in a colander until it stops dripping (don't press or squeeze it, though).
6. Lay meat on racks and set dehydrator to 160 F. Dry for 4-6 hours. Check the meat periodically. When it's done, it should be firm, but not brittle. It should bend, then crack, but not crumble or fully break apart.
7. Cool to room temperature and store in the fridge or a cool place (ideally vacuum sealed if it's not going to be eaten soon).

SAGE BREAKFAST SAUSAGE

Grinding up the smaller cuts to make sausage or ground meat (mince) is a fast and easy way to process many of the cuts and trims from goat meat. If you want a fattier sausage, include some fat from the harvest. Some people add pork fat, but there's no reason to not use the goat's own surplus! To make sausage and burgers, you will need a sausage grinder. Fortunately, you can buy very affordable hand grinders or spend a bit more for an attachment to your stand mixer (if you have one). Alternately, a dedicated grinder can be purchased.

1 pound (450 gm) fatty ground goat meat

1 tablespoon brown sugar (13 gm) or honey (21 gm), adjust to preference

3/4 teaspoon (1.5 gm) salt

1/2 teaspoon (1 gm) coarse ground black pepper

2 tablespoons (14 gm) fresh sage leaf chopped or 1 teaspoon dried ground sage

1/4 teaspoon (.5 gm) dried thyme

1/4 teaspoon (1 gm) dried ground garlic or 1-2 cloves fresh garlic minced fine

1. Mix all ingredients in a bowl.
2. Cover and let rest in the refrigerator for 30 minutes or more.
3. Form into patties. These can be cooked and eaten immediately or frozen for use later.

Part II   Meat

GOAT TALLOW

A home harvest of a younger, healthy goat will yield a decent to large amount of abdominal fat. When animal fat is cooked to remove its impurities, it creates a valuable product. At the time of writing, goat tallow online is over $30 a pint (16 oz)! Use goat tallow as you would beef lard for cooking, pastries, and even soap making.

1. Chop fresh goat fat into chunks and place into a heavy-bottomed pot. Place on low heat, stirring occasionally (A crock pot works well for small amounts). For the whitest, most neutral tasting fat, you want to cook it slow and low. If there's any browning, the resulting product will be stronger in flavor (which might be preferred!).

2. As it cooks, water will evaporate off, and solids (like bits of meat and gristle that might be present) will sink to the bottom of the pan. These solids might brown, but that's okay! When skimmed out, they're called *cracklings* and can be added to scrambled eggs and other dishes, although I've heard mixed reviews as to their deliciousness…

3. Once completely melted, place a sieve lined with muslin or a double layer of cheesecloth over a canning jar and carefully strain the hot fat into the jar. Some people strain it twice. It's important that the oil be perfectly clear. Any remaining solids will lead to faster spoilage.

4. Store the tallow in a cool place or refrigerator. It should be good for several months.

Part III
Fiber and Hides

Irresistible skeins of dyed natural goat fiber yarn at a spinning event in Newport, Oregon

Chapter Eleven
The Ideal Fiber Goat

Some of the world's most luxurious garments are made from the hair of goats. Not just any goats, but those selected and bred to produce long, soft hair that can be spun into strong, supple strands. There are two types of goat hair used as fiber: mohair and cashmere. I'll go more into this in a moment, but suffice it to say there's a lot of overlap and confusion when it comes to terms used in the animal fiber industry. For example, there are Angora rabbits (which include several breeds) and Angora goats (a single breed). Angora rabbits produce *angora* (sometimes inaccurately called *wool*) while Angora goats produce *mohair*. An angora sweater is made from the rabbit's fiber (angora wool), a mohair sweater from the fiber of angora goats, and, lastly, a cashmere sweater from the undercoat fiber (cashmere) of various breeds of goats. See what I mean?

Adorable Angora goats at the Pines Farm, Maple Valley, Washington.

Part III Fiber and Hides

When talking about goat fiber, one criteria of quality is its diameter, which is measured in microns. A micron is a metric system measurement. A single micron (μm) is one one-millionth of a meter. In other words: tiny. For a frame of reference more applicable to our daily lives, a hair plucked from our heads is usually between 17 and 181 microns and a faintly visible cobweb 3 and 8 microns. In order to be pliable and not scratchy, goat fiber must be fine and of low micron count. Goats (except mohair goats) have hair that is primarily made up of coarse hair called *guard hair*. In cold months goats, like most dogs, also

---

The ABCs of Fiber Type

In Nigora and Pygora goats (animals with both Angora and cashmere genetics), breeders might mention three fleece types. Each of these has qualities a breeder might prefer, so don't think of the letter ranking (A, B, and C) as if you were in high school! Here are the three types and a quick summary of what they describe.

A = Angora type, with silky mohair fiber about 6 inches (15 cm) long.

B = Blend of cashmere and Angora mohair about 3-6 inches (7-15 cm) long with a crimped texture.

C = Cashmere type, growing a fine undercoat about 1-3 inches (2.5-7 cm) long.

There are differences not just in the fiber itself, but in how that leads to herd management decisions. For example, type-A coats are shorn, type-B coats are shorn, combed, or plucked, and type-C coats are shorn or combed. Typically, in these hybrid fiber breeds, if the coat is to be assigned a ranking, it needs to be typed just before one year of age.

The Ideal Fiber Goat

grow a downy undercoat. The undercoat is called *cashmere*. Breeds that produce abundant and long cashmere are called *cashmere goats*. Their fiber, harvested in the spring by pulling, combing, or shearing, is the most valuable, but is also produced in the lowest quantities. Angora goats produce long silky fiber called *mohair* that is quite a bit thicker and longer than cashmere, but is still a wonderful fiber for fabrics. Mohair is produced in large quantities, compared with cashmere, and is harvested twice a year by shearing.

Angora goats are one of the oldest types of domestic goats noted in historical documents and images. The goat's origin in the mild, dry climate of Turkey and its dependence upon its thick coat for protection make it vulnerable to harsh weather when young (before their full coat comes in) and after shearing. The breed is also slightly less prolific, having more single births than other breeds. Breeders that focus on developing their herd for hardiness and health report increased health and fertility.

Fiber breeds include the noble Angora, cashmere-type goats, and crosses of Angora goats with cashmere-producing goats. The fiber from crosses is often referred to as *cashgora*. Goat fiber is free of lanolin, the waxy substance that protects a sheep's wool but is also an allergen for some people. Mohair is coated with its own protective substance, however, called *yolk*. The yolk keeps the mohair conditioned while on the goat, but must be removed by washing the fleece in very hot water before spinning.

In data collected by the United States Department of Agriculture, only Angora goats are tallied when fiber is addressed. It is likely that many cashgora- and cashmere-type goats are tallied in the "meat goat and other" category which comprises 80% of the total goat population. So even though the number of Angora goats in the United States continues to drop, from my observations, I believe that there are many small farms utilizing other fiber goats, such as Pygoras and even Nigerian Dwarfs.

Part III   Fiber and Hides

Fiber Terms

Break: A flaw in the individual hair caused by a sudden change in nutrition or stress.

Bobbin: The part of the spinning wheel that holds the spun yarn.

Carder: Hand or mechanical brushes with finely spaced steel teeth for smoothing and aligning fibers for spinning.

Carding: A mechanical or hand process that aligns fibers parallel with each other in preparation for spinning.

Combing: The use of steel-toothed combs to pull cashmere from the goat.

Crimp: The natural waves in the hair.

Crutch: The area on the animal between the back legs and up around the tail.

Crutchings: Fiber shorn from the crutch area, often stained or otherwise damaged.

Dehairing: The removal by hand or machine of guard hairs from cashmere-type fleece.

Draft: Separating fiber during spinning to control the thickness of the yarn.

Drop-spindle: A basic hand-spinning tool.

Felting: A process (purposeful or accidental) in which the fibers become closely tangled and locked together in a compact mass or sheet.

Hank: A measure of worsted yarn.

Kemp: Brittle, medullated, weak fiber. Considered an impurity.

Lock: A small, finger-sized bit of fleece that tends to stay together.

Luster: The natural sheen of the fiber.

Medullated: Fiber that has a hollow core or pockets of air, making it less soft, less lustrous, and not conducive to dyeing.

Micron: A measurement of diameter in units a millionth of a meter.

Nap: The blunt ends of fiber rubbed up or raised from yarn or cloth.

The Ideal Fiber Goat

Niddy-noddy: A tool for winding yarn into pre-measured lengths, or skeins.

Noils: Short, tangled bits usually removed from fiber during combing.

Pelt: The skin of an animal with the fleece still attached.

Ply: Strands of yarn twisted together during spinning. Each strand is one-ply.

Rovings: Long, loose collections of fibers after cleaning and carding that are ready to spin.

S-twist: Yarn spun counterclockwise.

Scouring: The separation of dirt and other contaminants from the fiber. Usually done using warm water and alkaline soap. (Many home processors use Dawn dish soap.)

Shearing: The removal, with scissors or electric clippers, of hair from the animal.

Shrinkage: The weight lost after fiber is cleaned.

Staple: The length of individual fibers in a roving.

Swift: A tool for turning skeins into balls of yarn.

Worsted: A yarn measurement of medium weight when compared to other thicknesses. The most common weight used by knitters and crocheters.

Wraps per inch (WPI): A way to measure the diameter of hand-spun yarn. Utilizes a WPI tool on which the yarn is wrapped.

Z-twist: Yarn spun clockwise.

Lice Issues

All goats are subject to lice infestations, particularly in the very early spring. The lack of sunlight and the realities of sheltering in tight bunches in bedding provides the kinds of conditions that lice prefer. In most goats, lice are a nuisance at best, and a threat to their health at worst. Mild cases will wane once the goats get out into the sun, but

Part III   Fiber and Hides

imagine the damage to the Angora goat's coat from the endless need to scratch! Ensuring the goats go through winter in the best possible health, with a high nutrition level and low stress, is the first defense. During shearing, inspect the skin of each animal so as to not pass an infestation directly to the next goat via the clippers. Then, if all else fails, there are several effective insecticides that can be employed.

What to Look for in a Fiber Goat

It's not surprising that the primary trait you look for when choosing fiber goats is the quality of their fleece. Other important things to assess are the goat's health, genetics, and general appearance—the strength of its structure such as feet, legs, and back. As a part of their health assessment, investigate available information regarding their fertility, size, and weight, and in the case of Angora goats, investigate signs that they might not be purebred (if that's important to you).

Fleece quality is assessed differently in goats for mohair and for cashmere. Both will be difficult to analyze if they have been recently shorn. If the breeder has records of shearing and yields, those can be extremely helpful in choosing breeding stock. Here's a quick overview of what to look for, but I highly recommend further research, including referring to the resources in the back of this book, reaching out to the breed/type organization you are interested in, and talking to breeders. There are several great books out on raising Angora goats, but not many on cashmere types. In fact, I found one that just came out, but it has a sheep on the cover!

With cashmere goats, look for the following:

1. Coarse guard hair, which makes these unwanted fibers easier to separate from the cashmere after harvesting.
2. A winter undercoat that is fine, crimpy, long, and without shine.

The Ideal Fiber Goat

With Angora (mohair) goats, look for the following:

1. Uniformity of fiber type, quality of individual fibers, and density of covering. Mohair type should be uniform throughout the goat's fleece. For example, there shouldn't be a mix of wavy and ringlet types.
2. Individual fibers should be glossy and soft. Avoid choosing animals with short, wool-like hair or dull, spiky hair.
3. Density of fleece is best evaluated by inspecting the neck and shoulders where 25% of the coverage will grow. If the animal has been shorn, loose neck skin, usually seen as many wrinkles in the neck, is an indicator that there will be a good deal of fleece growth—there's simply more skin.

The qualities that indicate a good general appearance are the same for dairy goats (discussed in depth in the first part of the book). All goats benefit from strong legs, feet, backs, and other hardworking parts. If the goat has been recently shorn, general appearance will be easier to evaluate. If not, be sure to put your hands on the animal and feel these areas as well as watch them while the animal is on the move.

Fertility, as I mentioned earlier, is important but easy to overlook when the animal is wearing a beautiful fleece. Interestingly, some sources suggest that high fertility can have a somewhat negative effect on fiber, at least according to the popular data. With high nutrition comes a slight increase in hair diameter, thus lessening its quality. It's important to note, however, that an increase in nutrition will increase fertility and birth rate, and since kid fiber is of the highest value, it makes sense to focus on that, rather than the quality of the adult fiber. (See sidebar for more)

Part III Fiber and Hides

Large goats might make more fiber, but in the case of Angoras, large size might be an indicator that they have been crossbred at some point. Besides oversized goats, look for horn growth that isn't the typical Angora pattern of low, swept-back horns if having purebreds is important to you.

> **Feeding for Future Fiber**
>
> Improved nutrition for the mohair-producing Angora goat is an important part of increasing your overall profits. Although the mother's own coat may show a decrease in fiber quality, it will pay off in several ways.
>
> First, the better the dam's nutrition at breeding, the more eggs are likely to be ovulated, thus increasing the number of fetuses, and therefore, the number of fleeces with the highest value (the kid fleece). Second, according to Smith and Sherman in Goat Medicine, the nutrition the fetus receives optimizes the number of hair follicles present in their coat. The better the nutrition, the denser the coat. Third, the better the dam's vigor during pregnancy, the less likely she will abort or struggle carrying the kids to term. And finally, the better the dam's nutrition after kidding, the more likely she will be able to properly nourish her litter.

Chapter Twelve

Fiber Goat Breeds

Even fewer breeds are dedicated to the production of fiber than those for meat or milk. Most fiber goats are also used for meat and some for milk: another example of the amazing versatility of the species! Let's go over some of the better-known breeds in the States and a few other places.

A registered cashmere doe, named EMW Ad Astra, exhibiting very long guard hair (hiding her thick undercoat).. Photo courtesy Danielle Choniere, Avalon Farms, Vermont.

Part III   Fiber and Hides

Angora

Origin: Area of Asia Minor that is now known as Turkey and in and around the district of Angora (now the capital Ankara). The breed was documented as early as 1571-1451 BC, and mohair is mentioned as far back as the time of Moses. In the United States, the American Angora Goat Breeders Association formed in 1900. Two registries for colored stock (as opposed to white goats) currently have websites, one formed in 1999 and one in 2002. Mohair production is currently the highest in the United States, Turkey, and South Africa.

Size: Bucks 180-225 pounds (82-102 kg). Does 70-110 pounds (32-50 kg).

Distinguishing Characteristics: Erect carriage of head; low, swept-back horns; long ringlets or flat silky white coat. As mentioned earlier, colored varieties have a separate registry started from colored Angoras culled from the original registry.

Fiber Production: Shorn twice a year for a total yield of about 10 pounds (4.5 kg) of mohair 5-6 inches (12-15 cm) in length. Fiber rated by type with that from younger animals being the most valuable. Mohair that grows in a preferred ringlet pattern referred to as type C. Flat growth is called *type B*.

Cashmere-Type

Origin: Goats selected for their natural ability to produce high-quality cashmere and then bred for that purpose can be called *cashmere goats*. While all goats (except Angoras) produce some cashmere, certain breeds do it better, particularly those from colder climates. Breeds in the rugged Kashmir region of northern India have long been used by the native population to produce high-quality fiber for fine garments. Today, cashmere-type goats in the United States are likely to have genetics that include Spanish meat goats from the Southern United States and feral goats from Australia, Tasmania, and New Zealand. The Cashmere Goat

Association (United States) was formed in 1992 (originally as the Eastern Cashmere Association).

Size: Varies, but typically a heavily muscled goat with a medium-to-tall stature.

Distinguishing Characteristics: To be productive, the length of the goat's cashmere must be at least 1 1/4 inches (3 cm) long with a diameter of less than 19 microns. Cashmere must have a crimped texture and is not lustrous like mohair.

Fiber Production: 3-8 oz (85-227 gm) a year per animal.

Heritage Navajo Angora

Origin: Developed in the American Southwest by the Diné of the Navajo Nation. These goats are quite similar in look to their Angora progenitor, for the Diné blended these with Spanish goats to develop goats perfect for the Four Corners part of the Southwest (where Utah, Colorado, Arizona, and New Mexico meet). Considered a multipurpose breed for milk, meat, and fiber. As with many things to do with Native American culture, the Navajo Angora has a convoluted history impacted by the US government's attempts to control Indigenous peoples and their land.

Kashmir Goats of Asia

There are many cashmere goat breeds in the parts of Asia where cashmere developed. Here are a few exotic-sounding varieties as listed in the online magazine agronomag.com.

- Changthangi (China and surrounding countries)
- Hexi (high-altitude, desert-adapted)
- Inner Mongolia Cashmere goat (Gobi desert region NE China)
- Licheng Daqing goat (dual-purpose)
- Liaoning Cashmere goat

Part III Fiber and Hides

- Tibetan Plateau goat
- Luliang Black goat
- Quzhumuqin goat
- Zalaa Jinst White goat
- Zhongwei cashmere goat

Nigora

Origin: This new breed's registry was founded in 2007 in the United States and is still developing foundation bloodlines. The breed is based on crosses between Nigerian Dwarf dairy goats (often good cashmere producers) and Angoras. Animals are intended for both milk and fleece production.

Size: Minimum height of 19 inches (48 cm) and a maximum height of 29 inches (74 cm).

Distinguishing Characteristics: No coat or eye-color restrictions. Small stature. Fleece designations identical to that of the Pygora goat (see sidebar "The ABCs of Fleece" page 102). The term "heavy" describes a Nigora that has a greater percentage of Angora than Nigerian Dwarf, usually having floppy ears and a type-A or type-B fleece. "Light" Nigoras have a greater percentage of Nigerian or miniature goat, usually having erect ears and type-C or type-B fleece.

Fiber Production: Same or similar to Pygoras below.

Pygora

Origin: First developed in the United States in the 1970s by Katherine Joregenson by crossing Pygmy and Angora goats. A breed association and registry formed in 1987 and currently has a small but supportive community of breeders.

Size: Does 65-75 pounds (29-34 kg). Bucks 75-95 pounds (34-43 kg). Doe minimum height 18 inches (46 cm). Buck minimum height 23 inches (58 cm). No maximum height for does or bucks.

Fiber Goat Breeds

Distinguishing Characteristics: The same colors allowed in Pygmy goats plus white.

Fiber Production: The amount of fleece a Pygora can produce depends on fleece type (type C produces the least and type A the most). Type A may produce as much as 3 lb (1.4 kg) of raw fleece per shearing while type C may produce only 8 oz (227 gm) of combed raw fiber (weight will be lost upon dehairing and cleaning). Type B will average somewhere in between.

Breeder Profile, Fiber Goats
Robin C. Oliver
Caney Fork Pygoras, Shade, Ohio

Robin Oliver and friends. Photo by Ben Siegel. (https://benwirtzsiegel.com/) courtesy of Robin Oliver.

"Goats make great sweaters," says Robin Oliver on her website. The veteran knitter should know, for it was her love of knitting that turned her from a fiber shopper into a producer of luxurious Pygora fiber. A former journalist with a background in marketing and raising horses, Robin took up knitting in her early twenties. She was writing a newspaper story about a group of women who donated hand-knitted hats to babies at a local children's hospital and was inspired to make a donation of her own. The knitters welcomed her to a Saturday morning social group at a local yarn shop for guidance on the project, and soon she had moved on from baby hats to scarfs, vests, and sweaters. As with most who hold knitting needles or a crochet hook in their hands and lovingly feed a strand of yarn into their painstaking craft, Robin was drawn to high-quality, natural fibers. But, as is also common, the price of such skeins of yarn was daunting. It was during her usual Saturday morning knitting circle when she learned the stunning fact that luxurious cashmere fiber came from goats. With that epiphany, she was hooked by the idea of someday having her own herd.

It would be more than ten years before Robin and her husband Wade would move from the city to rural land where her dream of owning cashmere goats had a realistic chance. But that didn't mean Robin didn't spend that decade learning about fiber types and the goats that produce them. When she stumbled upon a Pygora breeder in her area, Robin was introduced to small, adorable critters with the genetics of mohair-producing Angoras and cashmere-producing pygmy goats. Suffice it to say, it was love at first touch. Although originally Robin sought s producing a more cashmere-like type-C fleece (see page 102 for more on fleece types) over time she's come

to appreciate the typical Pygora type-B fleece. She said type-B fleece brings together the fineness of cashmere with the silky locks of mohair from the Angora heritage, making the fleece easier to maintain on the goat (being less prone to matting and collecting debris than the type-C fleece) while still offering next-to-skin softness.

Currently, Caney Fork Pygoras (named for the Olivers' first farm in North Carolina near the Caney Fork River) has a herd of twenty-two goats, sixteen of them does. At the time of writing, ten are bred and due to kid in the spring. Robin enjoys the close-knit (pun intended) community of active Pygora breeders, but at the same time encourages its growth through her marketing of not just fleece and fiber, but high-quality Pygora breeding stock.

Most of the farm's fiber is sent to mini fiber mills willing to process small batches of raw fleece into cleaned fleece and rovings. Robin loves blending Pygora fleece with alpaca and silk which she says creates a yarn with more memory, meaning knitted projects will not be as likely to stretch out of shape. Although Robin's marketing background has come in handy, she says she doesn't have to work too hard to sell the gorgeous fiber at fairs, festivals, and on her website. Best of all, she no longer has to visit yarn shops or check her budget before picking up a skein of decadently soft yarn for her latest knitting project.

For more, visit: https://caneyforkpygoras.com

Part III   Fiber and Hides

Chapter Thirteen

Harvesting Fiber

The fleece of fiber goats can be harvested in several ways. The most common method, and the only one that works for Angora goats, or crosses with type-A or type-B fleece, is to use clippers to shear it off. As I've mentioned before, cashmere-producing goats are best combed or plucked to remove the downy undercoat just before it is naturally being shed in the spring. Both methods are labor-intensive and not without their learning curve.

Danielle James Choiniere using a self-cleaning slicker brush to gather cashmere from one of her goats at Avalon Farms, Vermont. Image courtesy of Stephen Choiniere

Harvesting Fiber

Tools for the Fiber Harvest

- Shearers with 20-tooth comb head
- Blower (optional)
- Slicker brush
- Rake brush (for cashmere combing)
- Bags, paper or burlap, for fiber (plastic can be used for short term, but a breathable container will help preserve the fiber better)
- Pen for labeling
- Scale to weigh fiber
- Screening table
- Good lighting

Harvesting Cashmere

Cashmere-type goats are traditionally combed to remove the soft downy undercoat. This technique both takes advantage of the several-week period during which cashmere is ready to be harvested as well as minimizes the collection of guard hairs. Unlike shearing for mohair, which is done twice a year, combing for cashmere occurs only once. A goat's soft undercoat is grown from the summer solstice, when the daylight hours begin to shorten, to the winter solstice, when the days begin to lengthen. After that time, it is either shed naturally or harvested. Ideally collection begins before the onset of shedding. This maximizes the yield (as less cashmere will be shed into the environment), limits the amount of guard hairs collected, and limits damage to the fiber from scratching and rubbing as the goat tries to remove the itchy, no-longer-needed undercoat. An internet search will likely take you to some fascinating footage of these traditional methods in action.

Part III Fiber and Hides

Harvesting Mohair

Shearing is typically done every six months, fall and spring, to optimize the amount harvested over a year and the quality of the fiber. Several things must be considered to ensure the value of the fiber and the health of the goats before planning your shearing schedule. Although mohair from younger animals is the most valuable, it should be harvested the first time in the fall, after it's had a chance to lengthen and the kid has grown past the early months of life. Breeders often time the birth of mohair goats for March, planning for the first shearing in the fall when the kid is six months of age. Fiber harvested from older goats is often sorted by the location on the goat's body, as this correlates with quality. But even coarser mohair is of value for the manufacture of high-quality carpets, rugs, and upholstery. A particularly undesirable hair type that can be found in the coats of some Angora goats is *kemp*. Kemp is thicker in diameter than mohair and contains a hollow core. Fibers with a hollow center are called *medullated*. It's difficult to correctly evaluate for kemp without a microscope, but when compared side-by-side to white mohair, kemp will appear solid white, rather than translucent. This is because the core blocks the passage of light, whereas mohair allows light through.

Shearing Tips

Protection and shelter should be provided to limit stress and help keep the animals healthy. A few days before shearing, place the animals in a pen that is well sheltered and clean to help keep their fleeces clean and dry. Twelve hours before shearing, remove buckets or troughs of drinking water so their necks will be dry, as wet fleece is hard on clipper blades as well as slows the process.

The day of shearing, have ready paper or burlap bags and a labeling system. (Labeling is especially helpful if you're marketing the fiber.) A suggested labeling strategy is as follows: adults; kids, first shearing; yearlings (2nd and 3rd shearings); crossbreeds (if any); stained, dirty (with hay, bedding, etc.), and crutchings. Have a scale ready to weigh the fiber if it's important for your records (not just for marketing but for refining your herd's productivity genetics), a table to sort fiber from individual

Harvesting Fiber

fleeces (a table-top made of course screen is helpful as it will allow debris to fall from the fleece), a flat, smooth, and clean surface on which to shear, and good lighting.

Shearing is done with wide-blade clippers, the same type used by sheep shearers. A 20-tooth mohair blade is recommended by most Angora breeders. It's also advised to run the shearer at only 1500 rpms. A blower (such as those sold to prepare beef cattle for the show ring) can be used to remove debris from the fleece before shearing. Alternately, use a slicker brush to lightly comb the animal. Some shearers use a table and head restraint to shear the goat while it is standing, while others rock the goat back on its haunches or tie three of the goat's legs and roll it from side to side as it is shorn. When weather will be cool after shearing, an unclipped strip about 6-8 inches (16-20 cm) wide, called a *cape*, should be left along the goat's spine to help keep it warm and dry. Work quickly, but carefully, especially when near the udder so as not to nick a teat. There are several useful shearing videos on YouTube.

Marketing

According to the USDA's 2023 report on the sheep and goat industry in the United States, mohair averaged $6.35 per pound (2.2 kg). The range by state was extreme though, with a low of $1.10 and a high of $8.90—so you definitely don't want to count on the average price working for you. (This is one reason some producers send their fleeces to a consolidator to be sold.) Also, keep in mind that these are reported sales from those who complete the annual USDA goat and sheep survey. Even very small producers receive this survey. Who knows how thorough it really is. It's possible, if not likely, that there are many instances of small, local sales that are not reported. When trying to decide on pricing, I highly recommend reaching out to other producers in your region and online. Fiber festivals and events (a web search will reveal several) are a great way to take the pulse of this market—and meet like-minded flock folks!

Breeder Profile, Fiber Goats
Linda Fox and Bob Armstrong
Goat Knoll Farm Cashmere Goats, Dallas, Oregon

Linda Fox making sure the herd is friendly with tasty treats.

If there's one thing I should have been prepared for when visiting Goat Knoll Farm—and was not—it's the amazing fuzziness of cashmere goats in January! The variety of coat types in Linda Fox's established herd of almost forty cashmere goats was also intriguing, with some having tiny curls and others long silky strands of guard hair (the longer, coarser hair present year-round). But no matter the guard-hair type, the cashmere undercoat on each of Goat Knoll's animals was an abundance of luxurious fluff.

I arrived at the beginning of the fiber harvest season, the herd having been combed once. This process would be repeated every two weeks until the goat was freed of the insulating undercoat. By the end, each mature goat in Linda's herd averages 5-6 ounces of fiber. By the time the guard hair has been removed (*dehaired*) another ounce or so will have been lost. Still, (Continued next pg.) (Continued) the yield is impressive and a testament to Goat Knoll's careful selection since they got their first goats.

It started in 1995 when Linda and her then husband, Paul, acquired a small herd of three cashmere goats. But shortly thereafter, an opportunity presented itself that was "too good to pass up" when Linda and Paul found a herd of thirty-nine (that's right, thirty-nine) goats for sale just a few hours' drive away. At the time, they lived on a small lot, but already owned the fifty acres in Dallas, Oregon (in the coastal foothills west of Salem, the state's capital) that would become Goat Knoll Farm. Upon grasping the opportunity to acquire an established herd of cashmeres, the couple unanimously agreed it was time to move to the farm.

The herd's current quality reflects Linda's attention to statistical details. She keeps two fleece-related spreadsheets. The first is a combing worksheet with columns for each combing date and notes regarding timing and ease of harvest, goat temperament, body condition, and any issues that need to be monitored. The second is a fleece worksheet where she records her grading of fleece weight, fiber length, fiber diameter, and style score after the harvest is complete.

In addition to cashmere goats, a flock of twenty-eight Shetland sheep provide wool and meat. Both the sheep wool and the goat cashmere are sent to various small mills for processing into rovings and yarn. Linda hand-spins some of the cashmere—having the mills first remove the guard hair and process it into rovings—but most is sold to other spinners and knitters. She's experimented with a small dehairing machine (hoping to remove the guard hairs at home), but is so far unimpressed with the results. Much of the yarn Linda hand-dyes into vibrant, gorgeous colors. She also designs about one custom pattern a year and then creates kits for knitters using, of course, the farm's fiber and her own pattern, saying it both helps move the products and adds value to each skein of yarn.

I caught up with Linda and partner Bob Armstrong a second time, a few days later, at the Newport, Oregon Spin-Off, held in a school gymnasium and attracting about twenty vendors and a huge herd of spinners from the region. Popular events such as this occur enough times in the region to give Linda a great opportunity to market her fibers as well as meet up with other spinning enthusiasts. I was impressed at the number of attendees, some spinning, some shopping, and others knitting and crocheting in cozy circles in the center of the gym. Booths filled with gorgeous natural and dyed fibers, tools, and products lined the walls, none of which bore signs that said "Do Not Touch!" Which is a good thing, as it would likely be impossible to enforce.

I left Linda pedaling her oak Reeves Castle spinning wheel in lulls between eager crafters wanting to discuss Goat Knoll's stunning fibers and yarns. To order these luxurious, carefully crafted products or learn more about cashmere goats, visit http://fibergoat.com.

Linda Fox at the Goat Knoll Booth, Newport, Oregon Spin-In

Chapter Fourteen

Harvesting Hides

If you harvest goats for meat, you'll have access to a supply of hides, but if you don't there are some options. Lots of goat folks that harvest meat save the hides but never get around to tanning them. Stored in many a chest freezer are salted and rolled raw hides awaiting the producer's ambition. So, reach out to your goat community and you just might get a hold of a few.

A handbag I made from my first tanned hide. The hide was mineral tanned, and I used a home sewing machine and upholstery thread, along with a salvaged belt for the strap and ornamental buckle.

Part III   Fiber and Hides

If you've worked with other hides, say cattle or deer, you'll find goat skins to be thinner and more fragile—and the younger the animal, the more tender the hide. The term "handle with kid gloves" didn't start out as a metaphor! There's some advantage to this in that softening usually takes less work than working with a thicker hide, but the delicate skin will also be easier to damage.

For tips on skinning, please refer to Chapter 8, Harvesting Meat. The butchering section has a step-by-step guide. For this chapter, we'll start with a little background on how leather and hide tanning evolved and then get to work with the process, starting with after the hide has been removed.

Historical Tanning Technologies

The use of animal hides, mostly with the hair on (also called *furs*) to protect our early ancestors' feet and bodies, occurred far in the past, perhaps as soon as the first animals were hunted and skinned. If you've been a part of harvesting meat and seen the supple, fur-on hide separated from the animal, it's easy to imagine draping it across your own shoulders—if you were, let's say, previously unclad. The steps from this prone-to-rot garment to one of finished leather developed in different ways across the inhabited parts of the world. In other words, there's no single leather lineage, which means you now have several well-known technologies for making leather.

The different ways hides are treated and cured produce both leathers (hair-off) and furs (hair-on) suitable for different needs. All of these age-old methods can be considered "natural," with industrial chemical processing being relatively new (and not discussed here). Though the structural changes to the hide occur differently with each of these steps, all of them halt the decay process and the natural breakdown of the hide.

Harvesting Hides

Mineral Tanning (Tawing)
This ancient method (traced to around 800 BCE in Rome and Egypt) uses the mineral alum, or other aluminum salts, along with additional steps to make the resulting hide soft and pliable. Tawed hides can quickly be identified due to the white color of the leather. Many of the kits (including the ones I initially used) utilize this process. The biggest drawback to mineral-tanned leather is its vulnerability to water damage, both in staining and in stiffening. But for hides meant for interior use, this quick method is fantastic. If you want to keep the hair on, it's great because it doesn't dye the hair like vegetable tanning does. For a step-by-step guide, see page 137.

I would be remiss if I didn't mention the most common mineral tanning today: chrome tanning. This industrial process uses chromium salts. It has a large negative environmental footprint and is not utilized by the small-scale tanner.

Vegetable Tanning
Here's the method from which tanning gets its name. The medieval Latin word for *crushed oak bark* is *tannum*. Oak bark is high in tannins (therefore tannic acid), a natural chemical compound found in several plants and foods such as red wine and underripe persimmons. Tannins are astringent and bitter, but when it comes to converting hides into leather, they not only soak into the skin and bind with the collagen to make it stable (not likely to rot) and more water-resistant, but they also impart a lovely tan color. Tanning hides, hair-on or dehaired, using vegetable matter high in tannins is an ancient process. This method takes weeks to complete and results in sturdy, beautifully colored leather. In addition to oak, the bark of chestnut, larch, willow, sitka, white spruce, hemlock, and more have long been used for this process. I once tanned a small piece of hide in a madrone bark solution. The resulting goat leather was thick, on the stiff side, and a gorgeous mahogany color. Vegetable-tanned hides are followed with an oiling with tallow, lard, neat's-foot oil, diluted egg yolk, fish oil, or brain solution. For a step-by-step guide, see page 141.

Part III   Fiber and Hides

Fat (Brain) Tanning
When it comes to fat tanning, it's important to understand that all fat is not the same. Even the fat between muscles is different than that which is deposited around the organs and the fat padding our skin. Very few fats contain the right components to soften and help preserve animal hides. A brain is made of about 60% fat. In fact, it's, all other things being equal, the fattiest organ in the body! And most of this fat is of a type called *phospholipids*. Without going down this fascinating chemistry rabbit hole, it's phospholipids, particularly one called *lecithin*, that are effective for working with hides. Lecithin opens up the structure of the hide so that the other oils in the mixture can penetrate and soften it.

Another lecithin source often used by home tanners is egg yolks, which are also high in lecithin. I have not tried egg-yolk tanning, but I understand it is quite effective. I did try bottled liquid lecithin (from plant sources), which seemed to work quite well. It has the advantage of being shelf-stable and neutral in smell (not to mention cheaper than eggs, particularly at the time of writing this). You can also buy it in a bulk granular form.

Brain tanning is an effective method we can thank the indigenous peoples of North America for perfecting, although a few other cultures across the planet also used this method to preserve skins. It works best with the hair side (the hair and first layer of skin, the epidermis) removed so that the solution can quickly penetrate the entire hide before rot (putrefaction) sets in. In some, but not all cases, brain tanning is followed by smoke tanning. For a step-by-step guide to fat tanning, see page 145.

Harvesting Hides

Smoke (Aldehyde) Tanning

Smoking hides is an ancient method to treat the skin in a way that greatly improves its durability and usefulness. (It's preceded by fat tanning, but since smoking can be skipped, I wanted to give it its own section here.) Woodsmoke provides an abundant source of chemicals called *aldehydes*. (Woodsmoke can contain tannins if the burning wood does, but it's the aldehyde in smoke that does the work.) Modern industrial aldehyde tanning uses aldehydes in more concentrated, industrially friendly forms. Familiar aldehydes used in tanning include formaldehyde and acetaldehyde, although formaldehyde is no longer used (for obvious toxicity reasons—you might remember when it used to be in shampoos and all sorts of products).

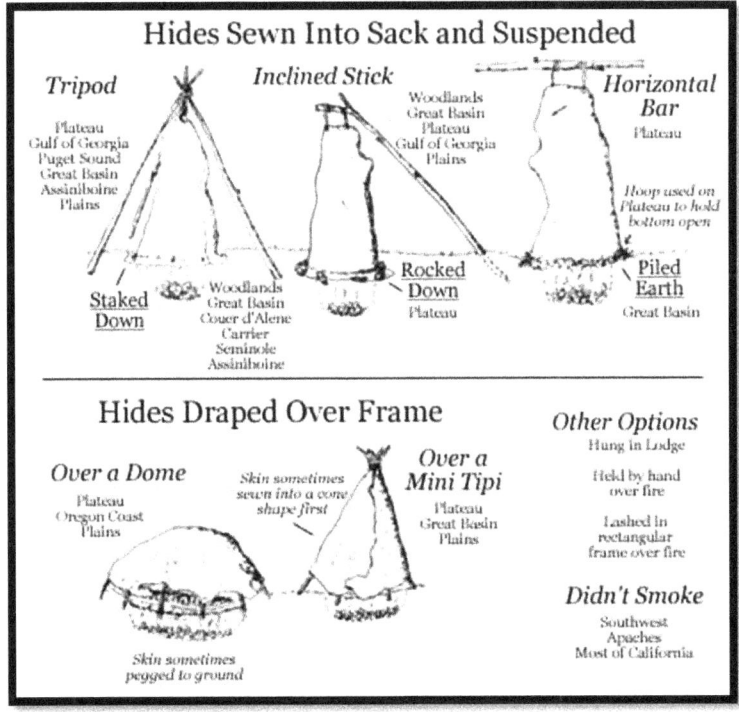

Traditional Indigenous North American Hide Smoking Methods .. Illustration courtesy of Matt Richards, Traditional Tanners. (See Resources for more information).

Part III   Fiber and Hides

Whether from smoke (in liquid or gas form) or other sources, these chemicals bind with the hide's collagen fibers, blocking their interaction in a way that keeps the hide supple as well as makes it more resistant to stiffening when wet. Smoke tanning is almost always preceded by some form of fat tanning and is followed by additional softening and working. For a step-by-step guide, see page 151.

Leather and Tanning Terms

Bating or puring (also *puering*): This is the softening of dehaired and scraped hides by exposing them to dung in a vat for a few days. The microbes and enzymes in the animal manure (often chicken, pigeon, or dog [called *pure*]) aid in breaking down the hides to a more supple state. Not surprisingly, this step is followed by a thorough cleaning. Today modern industrial tanning uses the protein-breaking enzyme papain (found in papaya fruit) for the bating step. The home tanner uses *pickling* to accomplish the same thing.

Bucking: The process in which the hide is soaked in wood ash or a different alkaline solution to remove the hair. Buckskins (the clothing made from soft leather) are made from dehaired, or bucked, leather from which the grain has also been scraped.

Buckskin: Soft, sueded leather (originally from buck-deer hides) used for clothing.

Chamois: A soft, highly absorbent leather named after the chamois, a European antelope, from which it was often made.

Fleshing: The scraping of the flesh side of the hide to remove fat, muscle, and membrane prior to tanning (more correctly: *defleshing*). A *fleshing beam* is a sturdy pole made of wood or plastic pipe, often braced against the worker's body or on its own rack, over which the hide is draped while being scraped.

Grain, full-grain: The top surface of a hide, the hair side is also called *the grain side*. Full-grain leather retains the pebbled pattern of the skin. In some leathers, the grain is scraped or buffed off.

Pickling: Soaking the defleshed hide in an acidic solution to dissolve nonstructural components of the hide and swell the hide in preparation for tanning.

Refleshing: Refers to repeating the fleshing process at some point to remove stubborn bits of membrane and flesh that were otherwise missed.

Rawhide: A hide that has been scraped clean of fat and inner membrane, then left to dry without any tanning steps.

Split-leather: Thicker hides are often cut (industrially) by a splitting machine along the thin plane of the hide, producing two or more sheets of leather. The bottom layer (formerly closest to the animal) is sometimes coated and embossed, or even bonded with a synthetic layer stamped to look like full-grain leather. It's buyer beware, as these products are often called real or genuine leather.

Suede: Often made from the lower layer after the hide is split with both remaining sides sanded and buffed into a soft nap. Suede can also be made from thin unsplit hides, such as goat.

Tanning Step-by-Step

Let me start by saying that if you're new to tanning and looking at various subreddits, forums, and product catalogs, IT'S VERY CONFUSING! Many terms are thrown about and used interchangeably, or even erroneously. There are tons of products designed to make the process easy, but NOT your understanding of what is happening *during* that process. I'm hoping to clarify some of that for you here. Some steps that follow are ubiquitous to all hide tanning processes; we'll start with those.

Part III   Fiber and Hides

Preparatory Steps

Scraping or Fleshing (Defleshing)

Please note, this step takes some practice! Goat hides, being thin and fragile, are tricky to learn on. Give yourself plenty of time and don't feel badly if you have to do a little mending of nicks.

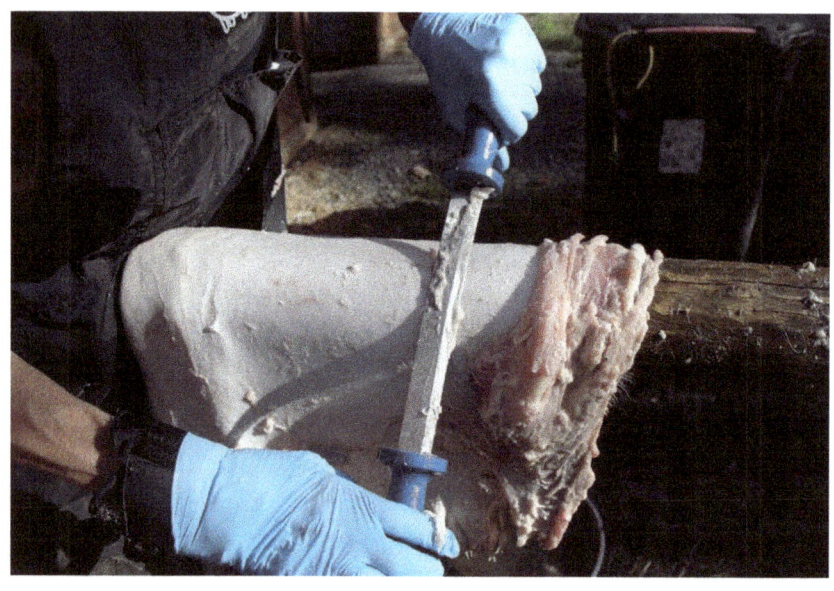

Scraping the hide is the laborious first step in any tanning process.

Hides can be scraped dry or wet. Dry hides have usually been stretched between a frame and dried quickly to prevent deterioration. Wet hides might be fresh or just thawed after being stored frozen. As part of preparing a previously frozen hide, immerse it in water for several hours. This helps plump the hide, as drying will have occurred in the freezer. If it was salted first, shake off the excess salt and lay the hide, skin side up, over a fleshing beam. Alternately, the hide can be laced or clamped snugly between a frame. You can make a fleshing beam from PVC or ABS pipe that is 4-8" in diameter (bigger is better) and cut to a length that you can brace on the ground and against your stomach, but watch out for notching the plastic with your fleshing knife. I used both

Harvesting Hides

ABS pipe and a wood hitching post (which left a few splinters in the hide).

Technique is critical when fleshing, as is patience as you learn. Work the knife away from you, holding the blade as close to flush with the hide as possible. Push all of the tissue away from you, moving across the hide laterally, rolling and peeling the fat and muscle tissue as you go. Keep a container handy for the scraps. If you have chickens free-ranging, they'll be more than happy to assist.

Washing (for Hair-on Hides)
Once the hide is mostly clean of bits of meat, fat, and membrane, you can hand-wash it several times to remove dirt and odors. I use Dawn dish soap and lukewarm water. Use as many washes and rinses as you need until the water comes clean.

A thick layer of salt being rubbed on the flesh side of a hair-on hide.

Part III Fiber and Hides

Salting (can be done as the first step)

Salting is a quick way to inhibit the bacteria that inhabit the hide and that, as soon as the animal is dead, begin to break down the hide and cause hair to fall out. If you want to keep the hair on, salting should not be skipped, but if you're dehairing the hide anyway, you can skip it. Salting should be done with pure salt—salt where the only ingredient is sodium chloride. In other words, no iodized or rock salt (like for making ice cream), as it contains impurities.

To salt, lay the hide, hair side down, on a sheet of clean plywood, a large garbage bag, or a sheet of plastic. Pour a thick layer of salt and spread it about ¼ inch deep over the entire flesh side. The salt will draw out moisture, so place the hide somewhere where the draining liquid won't be a problem. Let it set for 90 minutes to overnight. Check it midway, if possible, and add more salt to any areas where the salt has washed away. If you're salting before freezing, you can roll the salted hide and place the entire thing in a big plastic bag and freeze it that way.

Dehairing or Bucking: For Leather and Buckskin

An important thing to remember when working with hides is that the hair wants to come off! In fact, it's a battle to keep it on. Once the animal dies, the hair follicles holding each hair lose their grip on the hair and bacteria begin to grow rapidly in the follicle. **Hides you want to keep hair-on must be rapidly put into the tanning process** OR salted and/or frozen until the process can begin. The dehairing step is not necessary for all tanning methods, but for oil tanning (brain or other) it's very helpful to ensure a properly cured hide. When using ashes and/or calcium hydroxide or any other highly alkaline substance—as is used in dehairing—in addition to gloves and goggles, use a plastic tub rather than metal, which can corrode when exposed to the alkaline liquid.

Harvesting Hides

Plain Water: Because the hair of a hide "wants" to come off, a hide can be dehaired by simply immersing it in cold water. Historically, hides would be secured in the flow of a moving stream for a few days, being constantly flushed by fresh, cold water that would help wash away the hair. Check the hide daily, move it around in the water, and change the water if needed until the hair loosens.

hair-on hide soaking in wood ash taken straight (charcoal bits and all) from our woodburning stove.

Wood Ash: By adding ashes from a woodfire to water, you create an alkaline solution that very rapidly causes the hair follicles to release the hairs. This method is my first choice, because it is so effective and easy. Plus, if you have access to ashes from a woodburning stove or campfire, it's a free and easy source. The volume of ashes you need varies, but basically shoot for a mixture that looks opaque and has, when dissolved, a slippery feel on your fingers. It's not a bad idea to wear gloves, as anything too alkaline or too acidic is hard on our skin too! Mix the solution in a tub or five-gallon pail. Dehairing can be done at ambient temperatures, so don't worry about refrigeration during the rather messy process. Soak the scraped hide for a day and check for hair slippage. If the hair has not begun to fall off, add more ashes and/or give it more time. Be sure to agitate the solution and move the hide about multiple times during this stage.

When ready, the hair falls off easily. Here, I used a random piece of aluminum from a construction project to gently scrape the hair from the hide.

Lime: Calcium hydroxide, or hydrated lime, provides a powerful alkaline solution that works in the same manner as wood ash. Hydrated lime can be purchased from building supply stores. Be sure to wear a mask when working with this caustic, finely ground powder. (If you're wondering why this mineral is called *lime*, it's taken from the word *limestone*, from which it was originally mined. In turn, the word "limestone" comes from an old English word "lim," meaning "something sticky." Lime is the active ingredient in mortar for building rock walls and making cement. Fun fact, the word "slime" has similar roots!)Mix 1/4 pound (455 gm) hydrated lime with 2 gallons (8 l) warm water. Wear a mask and gloves when working with this solution! When the lime is in dust form, it's easy to inhale, and once water is added, it becomes caustic. The solution can be kept for some time. The above recipe makes enough for at least one hide.

Neutralizing: Once the hair falls off easily, the hide can be rescraped to remove the hair and the first layer of skin (the thin epidermis). You don't need the sharp side of your fleshing tool to accomplish this—so easily does the hair come off. The dehaired hide must then be brought back to a neutral pH. This can be done by repeated water rinses or rinsing first in water until it runs clear, then in a slightly acidic solution (usually using vinegar). If you're going to do that, a pH meter or strips is a good idea to ensure you don't change the pH too far in the wrong direction. As a reminder, 7.0 is pH neutral.

Pickling

Pickling, in any process, involves using acid to change and preserve. In the case of hides, it helps reduce bacteria and opens up the hide structure by dissolving much of the nonstructural tissue inside the hide, making it more porous and ready for the next steps. (In some traditional processes, where neither salt nor acid were employed, hides would be harvested, dehaired, scraped, then stretched and dry aged for several months.) If the hide is fresh, the pickling solution can be used as a storage solution (rather than freezing or drying), keeping the hide in great condition until you're ready to work it further. **This step is skipped by some tanners**, but it increases tanning quality and is usually worth the extra time. If you have plenty of hides and time, try it both ways and draw your own conclusions. Fun fact, in more primitive processes, urine, or bird or dog feces, were used for a similar step called *bating* or *puring*.

Hide-pickling solutions should ideally be brought to a low pH of 2.0. You can verify this with pH strips or a meter. In lieu of that, follow precise instructions on the product you're using. As with extremely alkaline solutions, please wear protective gear, gloves, goggles, and an apron, for this stage, and use a plastic tub rather than metal, which can corrode.

Part III Fiber and Hides

Vinegar (acetic acid) from the store varies in acidity. At 5% (check the label) vinegar is close to the right pH of 2.0—if you use it straight! At many building supply stores you can buy 30% vinegar (used for killing weeds and other strong acid needs) which is closer to a pH of 1.5. Now that doesn't sound like much of a difference, but thanks to the logarithmic nature of the pH scale (each full point difference is 10x more or less than its neighbor), it's a huge difference. The strong solution should be diluted to the correct pH as verified with pH strips or a pH meter.

Citric acid crystals (sometimes labeled *pickling crystals* or *pickle-crisp*) also work great. This product can be purchased from various sources, including cheesemaking, canning, and tanning supply stores.. The product is concentrated, so it can be stored easily and for long periods.

Other strong acids, like hydrochloric acid, work too, but for beginners, it's best to stick with easily accessible, less volatile ingredients.

Steps for Pickling the Hide: Dissolve 3 ounces (85 grams) of citric acid and add one pound (0.5 kg) of plain salt in every one gallon (4 l) of hot water used. For one goat skin, about 2 gallons (8 l) total is probably enough. Let this cool to room temperature. Submerge the hide and wait at least 72 hours, agitating a few times during the pickling. Check for readiness by squeezing the hide between your thumbnail and finger pad. When it's ready, the dent should remain (indicating you've plumped the hide fully). Test several areas, as some parts of the hide will take longer.

Neutralizing: Once the hide is properly plumped and passes the dent-test described above, rinse it multiple times to neutralize the acid. Some people add a bit of sodium bicarbonate (baking soda) at 1 ounce (28 gm) per 1 gallon (4 l) of water (making a mild alkaline solution) to bring down the pH. Soak the hide in this solution for 15-20 minutes, then remove and rinse at least two times.

Dispose of any remaining pickling solution by diluting and pouring on acid-loving plants (like evergreens and azaleas), adding sodium bicarbonate or a different alkaline solution to neutralize the pH and then pouring it down the drain, or pouring it on a driveway or gravel area where you want to make it hard for weeds to grow.

Refleshing (optional for leather and buckskin)
At this point, it's a good idea to lay the hide on the fleshing beam once more and, using the blade side or dull side, scrape away loosened bits of flesh and leached-out "nonstructural" constituents of the hide. This might appear as a powdery mush. The hide can also be thinned at this stage. It will be elastic and supple, so you can scrape thicker areas (usually the neck—aka the cape) and stretch the hide out to a more even thickness.

Mending
If holes larger than about 1/4 inch are present, you can stitch them up at this stage. You'll need a sharp, medium-sized needle (not too thick or too fine) and heavy upholstery thread or filament (fishing line filament works great, but is hard to manage when sewing). From the back of the hide, pull the edges of the hole together, creating a little ridge, and sew them shut. You'll have to be careful of these mends if you sand the hide later. Sewing is optional. If you are going to use the hide with the hair side up, and you cannot see the holes from that side, it is fine to leave them there.

Tanning Steps

Once the preliminary steps are done, the hide is ready to be transformed by tanning using one, or a combination, of the technologies we covered earlier.

Mineral Tanning (Tawing) Step-by-Step

If you buy a tanning kit, there's a good chance it uses aluminum sulphate as the tanning ingredient. A kit is a great place to start, as is mineral tanning, as it is quick, simple, and usually quite successful. You'll find a great variety of premixed products labeled in a bunch of different ways—rarely do they include an ingredient list! Be sure to read and follow

Part III   Fiber and Hides

whatever instructions are included with the product you choose. It's also easy to buy everything you'll need—salt, alum, and washing soda—individually and locally.

Mineral tanning can be done using a wet-method or a paste-method (most kits). I like the wet-method, it's easier to set up and work with, but if you don't have a good place to drip-drain the hide at the end, the paste method might be easier. Remember to wear gloves for all these steps! When mixing your own ingredients, know there are myriad instructions out there for the ratio of salt to alum and the amount—if any—of added washing soda. Please know that this simply underscores the many ways tanning can be accomplished. If you're more comfortable with a kit, go for it!

Mineral-tanned hides can be followed with a smoking step. This makes the hide more water-resistant and longer lasting. It will change its color, but can be done if you want to take your process a step further. (See the smoke tanning section on page 151)

Steps for Soaking Method

To make a soaking solution, mix 1 cup (287 gm) of salt, 1 pound (455 gm) of aluminum sulfate, and 4 oz (114 gm) of sodium carbonate (washing soda) in 2-3 gallons (8-12 l) of water. Immerse the hide in the solution, stirring well. Soak, stirring occasionally for 48 hours. Goat skins, being fairly thin, should absorb the solution well over that period of time. Check for completion by cutting into the hide (do it near an edge that you can later trim off). The hide should be even in color throughout. Remember, tawing—aka mineral tanning—creates a white-when-dry hide, so look for a light color. Hang the hide over a pole to drip for 20-30 minutes. When it is no longer dripping, it's ready for the oiling and softening stage.

Rubbing mineral tanning paste on the flesh side of a hair-on hide.

Steps Paste Method

For a paste, mix 1 cup (287 gm) of salt and 1 pound (455 gm) of aluminum sulfate with 1 cup (120 gm) of white flour, then add water until it's a spreadable consistency. Lay the damp hide, hair side down, out on a piece of plastic sheeting. If you want to be able to move the hide during the tanning process, place a piece of plywood beneath the plastic. Ideally the temperature during processing should be between 50-90 F (10-32 C). Wearing gloves, spread an even layer of paste over the skin side. Cover the hide with another sheet of plastic and weigh down the edges to keep the solution from drying. Let set for 48 hours. Check during this time to ensure the solution does not dry out. If it does, mist it with plain water. As with the soaking method, 48 hours is usually plenty for a goat hide. You can check for even tanning by cutting an edge of the hide and looking for an even color change. When done, scrape off the paste using a stainless steel or plastic kitchen spatula or scraper. (A pack of "plastic spreaders" designed for filling auto-body dents—aka Bondo scrapers—is quite useful for this stage.) Use a damp cloth to remove obvious residue. Don't rinse the hide. It's now ready for oiling and softening.

Oiling

Using a good-quality leather softening oil, such as neat's-foot or a proprietary blend such as Tanners Leather Oil, Liqua-Tan, or others. A lot of sites recommend mink oil, but I'm uncomfortable using a by-product of the mink fur farming industry. Whatever oil you use, apply it warmed and in a warm room. The oil should not be too hot to touch. Pour a generous amount on the hide, and wearing gloves, spread it out to the edges. If you're oiling leather (with the hair off) do both sides. Wait 2-4 hours, rub any residual oil in, then apply a second coat. Drape the hide over a board and allow it to dry for 2-4 days until it's mostly dry. You can check for readiness by using your fingers to stretch an area. When ready, the color will lighten from dark cream to light cream, and you'll feel the hide give. Now it's ready to soften.

Softening

In this step, you must stretch the hide so that the fibers of the skin, now leather, pull apart and become supple. If the hide is overly dry this is harder to do; when it's still a little moist, the fibers move more easily. A great way to work the hide is called *cabling*. Using a dog tying cable stretched tightly between two posts or even the edge of a sturdy table, pull and stretch every section of the hide across the cable or table edge. Don't stretch folded sections of the hide across the cable, or a semipermanent crease might be created! If you need to take an overnight break, store the hide in a plastic bag to keep it from drying too much.

Harvesting Hides

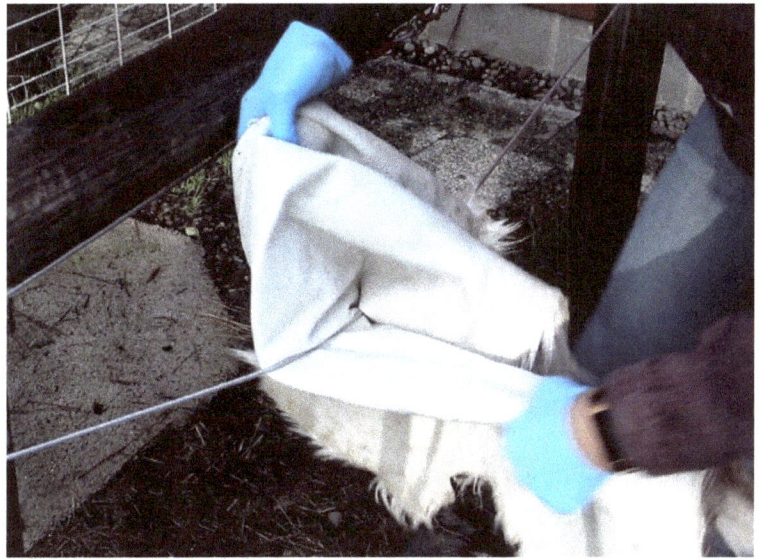

Using a coated dog cable to soften a hide.

Once the hide is supple and evenly soft, it's time to sand the surface. There will be lots of little bits sticking out and looking rough. Sanding takes care of this. Regular sandpaper works fine (as does a pumice stone). You don't need to by any specially labeled product. Start with a course grit and use a sanding block (also from the hardware store). Move to a finer grit as the hide becomes smoother. When it's to your liking, it's done! Remember, mineral-tanned hides are not waterproof. Water will stiffen them. But they can be softened again if needed.

Vegetable Taning Step-by-Step

As I mentioned earlier under the history of leather and hides, there are many plants that can supply bark or other matter with a tannin content perfect for tanning hides. Common options include oak and the bark of chestnut, larch, willow, sitka, white spruce, or hemlock. Non-bark sources such as acorns, leaves, and oak galls (known in my parts as *oak apples*) offer a source of tannins. Even bags of bark mulch from the

Part III Fiber and Hides

hardware store will work! Each source can produce different earth-toned colors. The important thing to remember when experimenting is that the plant matter should be gathered *before* it has been leached by rain or sprinklers—that's what *you* want to do when you soak it in water! Each source will vary in tannin content, but you will develop a feeling for it as you observe the bark tea, or liquor, as it is often called. Even dipping a finger in and tasting for bitterness and astringency can be helpful in understanding how strong the bark liquor is.

I only bark tanned once—successfully, that is! In my first attempt, the bark I used was too old and didn't make a strong enough tea. In the second, I used madrone, a native tree in our area that I adore for many reasons, one being its beautiful red bark. Not expecting success, I only tanned a small piece of hide, but it worked so well, I sure wish I'd done more. The hide thickened and turned a gorgeous mahogany.

Making the Bark Tea

Bark can be chopped from thicker logs and stripped with a draw knife or hatchet from young branches. Chop the bark into one-inch pieces, place in a stainless-steel pot, and add water. If you have hard water, meaning water with a high mineral content, try to use rainwater or other "soft" water to prevent minerals from reacting with the tannins and creating blemishes on the hide. Iron, in particular, from water or bits of blood on the hide causes black stains. In fact, some tanners test their bark liquor's strength by immersing an iron tool or implement in the liquid. If the tannin content is high, it will quite quickly turn the steel a deep black. Bring the water and bark to a boil and simmer until the bark tea is a rich brown color. Strain the bark out, saving it if you want to do a second batch, and cool. The tea should be a nice rich brown color. This same process can be accomplished without boiling the water, but it will take longer.

Harvesting Hides

Tanning the Hide

A note: Cow and buffalo hides can take 5-6 months to fully bark tan. Deer hides take 3-4 months. Goat hides are much thinner, so it might only be weeks until they are ready. Initially, prepare to check the process more frequently until you get a feel for what's working for you.

Bark tanning is best done in two stages. In the first, the bark tea should be a weaker solution. If it's too strong, tannins will fill the outside layers of the hide, blocking absorption to the center. Folks who tan frequently will take the tea from a previous batch (sometimes called a *spent liquor*) and use it for the first soak.

Tannins absorb into the hide best when the solution is between 45-85 F (7-29 C). Place the hide in the solution in a large plastic container, weighing the lid down with a rock or ceramic plate (don't use anything corrosive, like aluminum; even some stainless steel is too poorly made to work; in the old days, wooden or stone vats were used). Stir daily. Some people wring the hides every day to help work the solution into the hide. After a few days to a week, check for even color distribution on the surface of the hide. Pale patches on the surface mean there are still bits of membrane that need to be scraped off. Remove the hide and reflesh it. If it looks even and isn't too dark, progress to a stronger solution by adding a fresh batch of bark tea. **Basically, it's better to go too slowly than too quickly!**

After another week or so, make a small sliver into the hide (do this in an area you can trim off later) and check for even color distribution. It's unlikely to be ready, but it's important to monitor the progress as you get a feel for what's going on. Once it's even throughout, the tannins have reached the center and the process is done. Tannins thicken the hide. The longer you leave it in the solution, the heavier and less stretchy the hide will become. It is said that the volume of tannins a hide can absorb might add 50% to the weight of the original hide! Also remember, if you're bark tanning with the hair on, there will be color changes to the hair follicles too.

Part III   Fiber and Hides

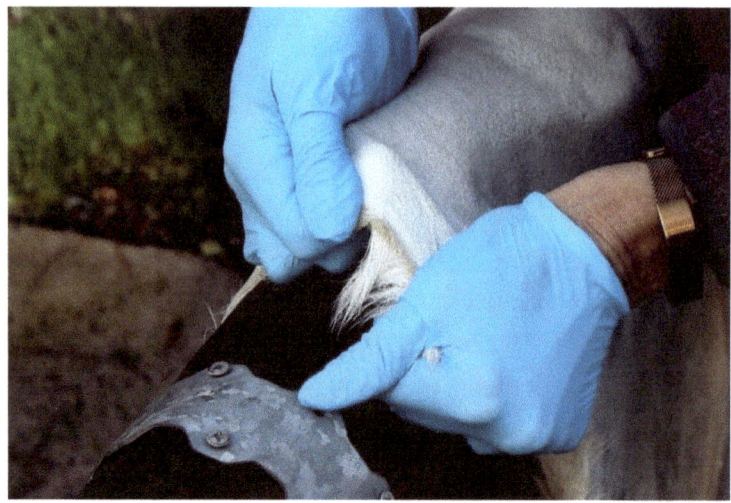

Hand stretching a slightly damp hide.

Softening and Finishing

As with all tanning, a step follows that softens the tanned hide to make it more usable. In the case of vegetable tanning, there are various ways to accomplish this, from repeatedly rolling the hide in both directions (grain side out and then grain side in) as it dries to scraping and stretching it on a flat surface. Some people begin working the hide while wet. Some wait for it to dry. When you start reading tanning forum posts and blogs and watching videos, you're going to see myriad ways to accomplish this step. What I'm sharing here, as with many of the tanning processes, is just the tip of the iceberg.

If the goal is to also flatten and smooth the hide into an even sheet (called *slicking*) then a tool called a *slicker* is extremely helpful. Slickers are made from glass (often with a wood grip) or slate. But a piece of ceramic tile will also work, as long as the edge is gently rounded. During slicking, the hide should be stretched in a frame or laid out on a smooth, clean surface, also known as a slicking table or board. A popular approach uses tallow or another thick oil to paste the flesh side of the still-damp hide to the slicking board, holding it in place. The hide is then worked with a slicker until smooth, then left on the board to dry.

Leather conditioning oils such as olive, neat's-foot, tallow, brains, and/or fish oil can be applied to either or both sides of the hide if more softening is desired. Oils will darken the color, something you may or may not want!

Fat Tanning Step-by-Step

I'm including three fat-tanning methods here: brain, egg-yolk, and lecithin. There are others! Fish oil has been used by indigenous cultures to tan hides and fish skins, and you'll read about modern tanners using neat's-foot oil (rendered from beef bones) and even soap with added oil. The takeaway is, this section is not exhaustive.

Regarding brain tanning, it is often said, "Every animal has exactly the right volume of brain to tan its own hide." This is true, but first you not only need to prepare the brain, you have to retrieve it from the goat's skull—if you did the butchering yourself. I found this the most daunting part of the process (see sidebar for instructions), partly because the skull is thick and hard to access and partly because working with the goat's severed head is emotionally hard—at least for me. Don't feel badly if you find it too personal and intimidating! You can always buy beef brains at a butcher or use egg yolks or ready-made lecithin.

Make the Tanning Solution

Brain Solution: Puree 1 brain per hide in a small amount of water, then mix into 1 gallon (4 l) of warm water (about 120 F [49 C]) and 3 tablespoons (44 ml) of oil (this can be rendered lard, olive oil, or other).

Egg Yolk Solution: Mix 10 egg yolks, 1 gallon (4 l) of warm water (about 120 F [49 C]), and 3 tablespoons (44 ml) of oil (this can be rendered lard, olive oil, or other).

Part III   Fiber and Hides

Lecithin and Oil: Combine 1 gallon (4 l) of warm water, 2 rounded tablespoons (30 gm) of lecithin crystals or liquid, and 3 tablespoons (44 ml) of oil (this can be rendered lard, olive oil, or other).

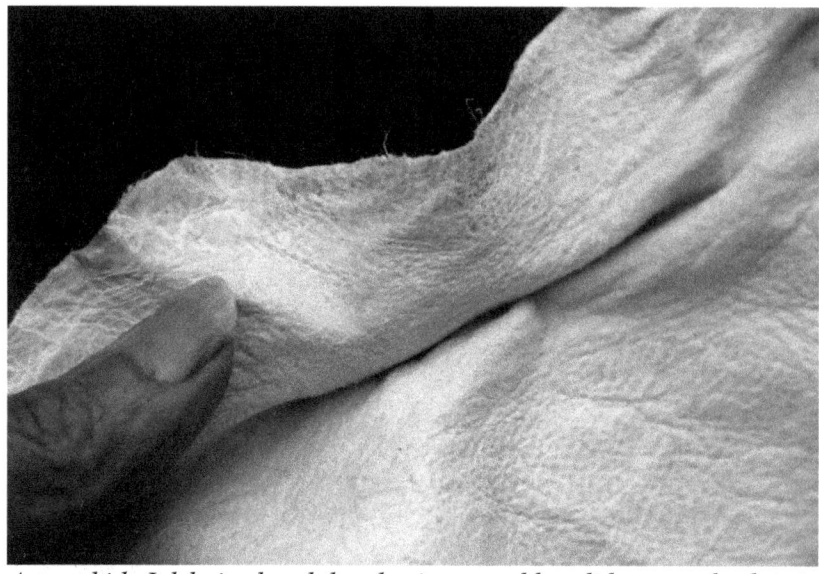

A goat hide I dehaired and then brain-tanned but did not smoke, leaving it soft, supple, and absorbent.

Soaking and Wringing

Immerse one hide in one of the above solutions for at least 15-20 minutes. Wring out using a twisting pole if available. This step helps open the structure of the hide so that the solution can fully penetrate. Soak again. Wring again. Do the entire soaking and wringing process at least 4 times.

Stretching and softening

The hide must now be worked until it is fully dry. Goat skins with the hair off don't take too long, unless the weather is cool and humid. The goal is to keep the hide moving and pulling in multiple directions. You can accomplish this in several ways. Two people can literally pull from opposite sides and work their way around the hide; you can use a taut cable or the edge of a table over which you "squeegee" the hide back and forth (see picture on page 141). A hide can also be stretched by lacing or clamping it to a sturdy frame and then working it with a dull blade, scraper, or bone tool.

Finishing

If you stop at this point, the hide will be fairly well preserved and very much like an absorbent chamois. But if you want to make the oil-tanned leather longer lasting and water-resistant, then smoking is the next step.

Smoke Tanning Step-by-Step

As mentioned earlier, smoke tanning is an additional process that comes after fat tanning or mineral tanning. This is the only tanning method I have not tried, but fortunately it is fairly straightforward and offers a wide variety of flexible approaches. (See image on page 127)

Different woods create different colors when smoke tanning. If you progress with your process, you might want to experiment with wood types. Smoke tanning can be done with hair off or on, but in both cases, the smoke is applied to the flesh side of the hide. In indigenous cultures, where an outdoor fire was readily available, hides were stretched between poles and then placed near the "cool" smoke. It's important that the hide doesn't get too hot but is slowly infused with smoke, allowing the aldehydes to infuse the protein matrix of the skin.

Part III   Fiber and Hides

> **Getting the Brain Out**
>
> Before working with the brains of certain species, make sure there is no risk of exposure to contagious "prion diseases," including chronic wasting disease (CWD) in the deer family; bovine spongiform encephalopathy (BSE) in cattle, and scrapie in goats and sheep. If you're certain your brain source is safe, then removing it from the skull is the next step.
>
> Usually this is done by removing a square plate of skull from the top of the animal's head. Using a short, sharp knife and mallet, or even a reciprocating saw, cut lines around the top of the skull where the horns are, or would be. If the animal is horned, it's more awkward. You might need someone to hold the skull (the horns are handy for that part) or a clamp of some kind. Cut just through the bone. Once released, remove the plate, exposing the brain. Scoop the brain matter out with a spoon, placing as much of it as you can get into a plastic bag or bowl. If you've killed the animal with a headshot, the brain will have some damage, but it doesn't seem to matter for tanning. The brain can be frozen in the bag until you need it.

Prepare the Hide
Loosely stitch, or clamp with plastic clamps, the hide (already fat or mineral tanned and softened) together on the long sides and one end,

Harvesting Hides

creating a bag. You can use big, fairly loose stitches. If the hide is hair on, leave it on the outside.

Some tanners create a fabric "chimney" to funnel smoke into the hide and keep the edges of the skin from touching the hot metal (when using a pail to hold the coals). Use denim or canvas. Old pairs of jeans work well, but might need to be sewn together to fit around both the bucket and the open end of the closed hide. Use a tight row of clothespins to secure the skirt to the open end of the hide.

Build the "Fire"

Prepare a small rock-ringed fire pit or a metal bucket to hold the coals. The top of the bucket or fire pit should be slightly smaller than the bottom opening of the hide bag or fabric chimney, if you use one. Lining the bottom of the pit or bucket with rocks helps keep the coals hot (allowing a little oxygen to surround them)—and in the case of the bucket, a few holes at the bottom will allow in a little oxygen to help keep the coals burning. If using a bucket, build a fire separately, then let it die down to a few good coals. Shovel a scoop of these into the bucket. On top of the coals place chunks of totally dry, mostly rotten (punky) wood. If you find the wood wants to burn too quickly, you can partially smother the fire by placing a metal plate (like a hubcap or the lid to a metal pot) into the bucket.

Part III   Fiber and Hides

As when smoking cheeses, hides need to have "cold" smoke. At Pholia Farm, we retrofitted a smoke with a soldering iron inserted into wood chips to provide cool smoke. Later, we drilled holes in a defunct refrigerator to make a large unit. Such a set up can also work for smoking goat hides (or other small hides).

Harvesting Hides

Alternately, hides can be smoked in an actual smokehouse or using elements of a meat smoker. If you're one of the rare folks with a smokehouse, then you can simply hang the hides inside. If you want to use elements from a small smoker, use an electric hot plate with a pan filled with wood chips or pellets as used for cold-smoking meats or cheeses. If the hide is small, you can even use a soldering iron in a can on its side (see image). Be sure the tip is covered with pellets or chips and check it periodically. Get it smoking, but don't let the chips burst into flames. Wetting them will slow the fire, BUT will steam the hide, which is not helpful at this point.

Smoke the Hide
When everything is ready, suspend the open end of the fabric skirt or hide over the pit or bucket using a line (rope or other) attached to the top of the hide and running up to a branch or beam. If large amounts of smoke are escaping the hide through any holds, plug them by applying a clip to hold the ends closed or pushing a chunk of punky wood into the opening. Check the smoke source every 20-30 minutes and, if needed, add more chips or rearrange the wood chunks. It can take only an hour to achieve an adequate smoking level. Check the hide for color and saturation. If the hair is off, it's easy to see the changes from the outside. Some tanners reverse a hair-off hide. If it's sewn into a bag, just turn it inside out and re-smoke it. Goat hides, being thin, might not need this. The neck area, though, is always thicker, so pay attention to that section. When you like the color, the smoking stage is finished.

Finishing
Undo the stitching and roll the hide up. Let it set for a day or two, then it's done!

Part III Fiber and Hides

Mountain Lodge Farm manager Brette Langhorn (left) and cheesemaker (right) Gorby Just showing off Gorby's first hide tanned.

Part IV
Pack, Cart, Brush, Pet, and More

*A "professional" brush goat hard at work.
Photo by Linda Williams.*

Chapter Fifteen

Pack and Cart Goats

Some of the most charming old photos are of goats pulling carts, often driven by well-heeled children in Victorian dress. One of my favorite books (now out of print and hard to find—but worth it), *The Mysterious Goat* by Dr. C. Naaktgeboren, is filled with not only delightful images, such as Victorian-era cart goats, but fascinating history. Goats as draft animals have served humans in many capacities, including by amusing the wealthy's youngsters, assisting with necessary work, and even acting as the "horsepower" for larger conveyances. One memorable example is that of the once-famous "Goat Man" of Georgia, Charles McCartney, whose large team of goats pulled his hodgepodge house on wheels around the American South for decades in the mid to late 1900s. (You can find lovely articles and photos of the Goat Man with a web search.)

Goats have pulled carts and been family companions for centuries. This Victorian era image makes the goat appear a noble beast for the elite. Photo from iStock. Credit Whitemay.

Part IV   Pack, Cart, Brush, Pet and More

Today goats serve as skilled and sure-footed pack goats for trips into the wilderness and the engines of pull carts as a novelty. Your goats might be full-time packers, pullers, brushers, or weekend warriors. Even milk goats might enjoy a little change of scenery; think of waking up in your tent to a hot, fresh goat's-milk latte—that is if someone else gets up and does the chores.

The Pack and Cart Goat

While in theory any goat can be trained to pull a cart or carry a packsaddle, if they're going to be "professionals," then their size, structure, and demeanor are important. Pack and cart goats are often dairy goat wethers, males that have been castrated, usually at an early age. This is both because of the ready availability of dairy goat males and the fact that castrated males, of all livestock species, grow taller than their intact counterparts (the energy that would have been used developing what are called *secondary sex characteristics* such as a thick neck and handsome beard are instead diverted to increased stature). These castrated males may end up at the same final body weight as their intact counterparts, but will have longer legs—not to mention a work ethic that overcomes the intact buck's powerful sex drive. Being larger and of different build, males are capable of pulling and carrying about 50% more than female goats of the same breed. Be sure to keep this in mind when deciding to put the effort into training a goat to be a draft animal.

Because pack and cart goats should begin their training very early in life, you won't be able to assess their structure thoroughly until they are older. Instead, look at the parents' general appearance (based on what I describe in Chapter 1). Focus should be placed on legs and pasterns that are strong and properly aligned. This doesn't mean the animal's toes can't point out just a little, or its toes can't splay a bit, as these traits can help the animal gain, and maintain, solid footing on uneven surfaces. If the toes do point out, then the pastern and knee should align with that conformation or there will be extra stress applied to the misaligned joints. (Note: On the front legs, the part we call the knee is actually more like our wrist, but calling it a knee is acceptable.) Moving up from the feet and knees, their shoulders and chest (sometimes called the *front-end assembly*) must be well built. The shoulder blades should fit snugly to the body, and the chest should be wide enough to accommodate a strong heart and lungs. (Note: As with humans, the shoulder blade is only held to the body by tissue. It is not an articulated joint. Any looseness of attachment will lead to issues over time.)

When buying a kid for cart or pack purposes, decide ahead of time if you plan on only having goats with or without horns, as the window of opportunity for proper disbudding (the removing of the nascent horn and the cells that grow it) usually occurs within a week of birth. Pack goats with horns are quite common, as horns help the goat dissipate heat while working and can be used for defense against small predators if needed. But if you have other goats without horns, or if you have fencing, and/or feeders where horned animals could get stuck, or any other safety concerns for the goat or people handling it, just know that goats can still be superior packers without them.

Equipment for Packing

When choosing equipment keep in mind the type of packing you'll be doing. For example, if you only plan on doing short day trips, then you can choose lighter duty equipment such as soft, all-in-one day packs. For long overnight packing you'll need a well-fitting packsaddle, with rigging and pannier bags (think saddlebags). The packsaddle and pannier are designed for heavier loads. The saddle and its pad spread the weight over the goat's back. When fitted properly, and not overloaded, there should be no discomfort to a goat who is trained and in good condition. As a rule of thumb, a well-prepared pack goat wearing a properly fitted saddle can carry 25-30% of its body weight.

Part IV   Pack, Cart, Brush, Pet and More

Panniers and rigging for light packing from packgoats.com. Image courstesy of Marc Warnke and packgoats.com.

A collar or halter and leash are used to guide the goat behind you on the trail and to tie them. Often, ground ties are used to secure a string of goats at resting and camp sites.

Panniers, the saddlebag part of the gear, are usually made of a tough, waterproof material and come in different sizes and configurations of pockets for organizing gear. A day pack, as mentioned before, is basically a pannier with straps that secure it directly to the animal. They're also of use when training the goat to pack before they graduate to the saddle.

Pack and Cart Goats

Packsaddle frames might be made from wood or aluminum. A saddle pad goes under the frame of the saddle to provide comfort for the goat and help the saddle fit well. The rigging consists of a strap, called a *cinch*, that goes around the goat's chest, or heart girth, just behind the front legs; a strap that goes around their chest; and one that goes around their rump. When all of these are properly fitted, the saddle should stay in place—no matter the terrain.

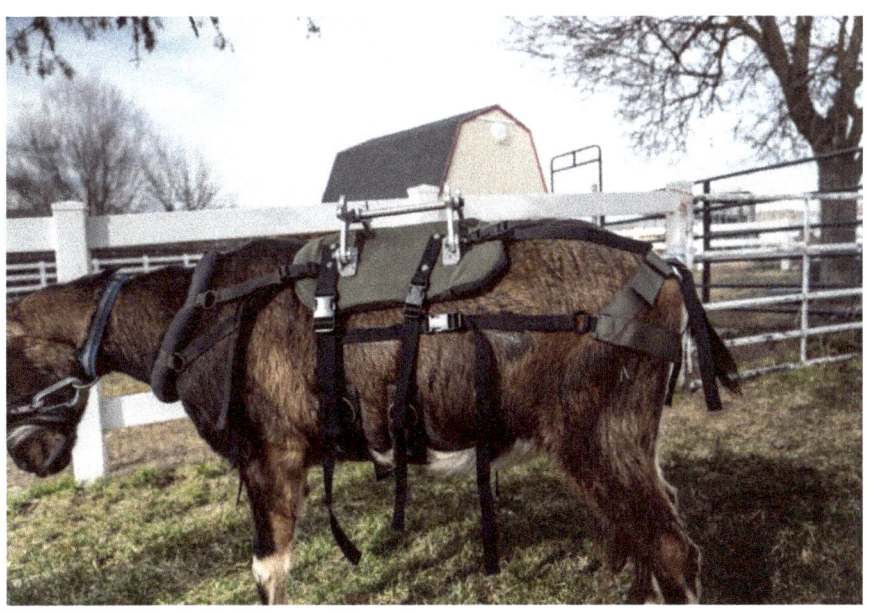

Goat packsaddle (without a load) from packgoats.com. Image courtesy of Marc Warnke and packgoats.com.

Breeder Profile, Pack Goats

Marc Warnke

Packgoats.com, Top End Adventures, and Ripple Ranch, Boise, Idaho

Marc and some of his goats on a backcountry adventure.

I found Marc, aka "the Goat Guy," when searching for companies selling pack goat equipment. I was thrilled to see such beautiful, thoughtfully designed packs and saddles on the packgoats.com website. Initially, I reached out simply to request permission and copies of photographs to use in the book, but the more I investigated the site and Marc, the more I realized that what he's done—and is doing—would be a great profile addition to this book. Fortunately, Marc was amenable to the same!

Based in Boise, Idaho, Marc's operation includes not just the sale of pack goat equipment, but backland tours, training through videos and live workshops, the sale of goat kids with proven pack genetics, and a generous serving of philosophy. Basically, packgoats.com is a gateway to experiencing the wild, goats, and life in, what is for many, a whole new way. Located where they are in Idaho, the company takes full advantage of the abundant public land and extensive trail systems in the state, and goats are key to this utilization, being able to transit the myriad ungroomed trails and remote areas where horses, donkeys, and mules cannot—and mechanized vehicles cannot and/or should not.

Marc got his start with goats in 2012 with a doe named Daisy, who Marc says " was probably an Alpine." It wasn't the breed that impressed him, but the goat's personality and potential. He soon realized he could combine his new fondness for goats with his passion for wilderness adventuring. But his first experience with the goat-packing equipment that was then available was disappointing. So, through trial and error he developed the saddles and packs and gear he now sells, trail testing it over at least 500 miles of real backwoods work. "Four or five years later," Marc says, "the business was born."

Already a public figure in the hunting world, it didn't take Marc long to integrate high-end backcountry (Continued next pg.) (Continued) adventures into the business. These trips, using his goats, of course, to pack supplies in and out of the wilderness, have brought guests from all over the world. With each trip, he helps elevate goats from the "200 years of misunderstanding" they've experienced on this continent. (Be sure to check out Tami Parr's book, *Goats in America: A Cultural History*, if you want a deep dive into this remarkable story.)

I asked Marc share the most common misconceptions—and the reality behind each—that he faces when teaching prospective caprine owners about working with pack goats. "The first is that they are easy to raise," Marc says, citing what experienced goat owners know well, but folks who've only seen pictures of goats (Continued next pg.) frolicking on cars and thriving (purportedly) on tin cans, do not. "The second is the importance of good genetics." Like with other working goats, such as dairy animals, only through careful breeding over time can you have more consistent offspring that are both mentally and physically suited to their jobs. The third is the importance of a good packsaddle. "If it doesn't fit well and isn't built well, the goat won't be able to carry its load comfortably—and the saddle won't survive the rigors of remote work," Marc states. And lastly, "The importance of conditioning—both the animals and yourself—to walking long distances, carrying loads, and adapting to changes in altitude." Marc and his team seek to overcome all these misconceptions and more through their goat mastery classes online and in person.

I asked Marc how he sees himself years from now, when the nine-day backwoods adventures he loves might be a little too much. "I'm a pretty focused guy," he said. "I hope, even into my eighties, to be out there exploring: hunting for rocks, crystal caves, and maybe even still an elk. Most of all," Marc added at the end," I want to keep helping people."

Learn more about Marc and his pack goats at https://packgoats.com

Pack and Cart Goats

Training the Pack Goat

I have trained quite a few horses to be ridden, starting them on the ground and slowly graduating them to carrying flesh-and-blood cargo. Training goats to pack is quite similar, but without the risk of getting bucked off! A good pack goat should be handled from birth and bonded with humans. For that reason, it's a good idea to raise them on a bottle, or at least handle them daily in a way that benefits the young goat. Once bonded, the training begins with teaching them to tie and lead, then to follow you in all types of conditions, and finally, when they are old enough, to carry a load.

Goats can be trained to lead at a very early age. I suggest at about eight weeks old, by first clip-tying them with a short, double-ended snap attached to their collar (made of webbing) and then to a fence. Stay close when they are first clipped as they may decide to try to twist their way out of it and end up tightening their collar. It only takes a few sessions for them to understand the limits of the clip and remain quiet. From that point you can extend their sessions so that they learn to stand patiently. Once this happens you can add a length of rope or webbing to accustom them to being tied that way. Again, stay close, as at first they are likely to become entangled. Unlike most horses, however, goats will quickly learn how to untangle themselves and soon how to avoid it altogether. During this training, you can teach them to follow you on the lead or leash. A goat's instinct is to follow their herd, so as long as they see you as their herd leader, this is the easiest part of their training.

When your goat is good at tying and leading you can start exposing them to conditions they may encounter when packing. Car noises, dogs, water obstacles such as streams and rain, all of these things and more might be encountered on the trail. It's critical they be accepting of these things before they are carrying a load in situations where their fear might result in disaster. Rain and water features are one of the hardest things for the goat to accept—as goat people know, goats seem to think they will melt if exposed to water. Take your goat out in the rain on pleasant adventures or lead them through it to a feed treat. Anything you can do to help them overcome their natural instincts will make for a happy

Part IV   Pack, Cart, Brush, Pet and More

relationship when packing. Regarding exposure to loud noises, such as gunshots if hunting, cars, and such, introduce these from a distance. Goats are smart and will soon learn that no harm comes of these noises.

A string of well-trained pack goats will wait patiently at an exhibition or an overnight hike

When your goat is long past weaning and closer to a year old, you can use a small day pack to begin saddle training. Taking it on and off will accustom them to the straps. Do it in short increments with no weight and little pressure. Very soon they will be accepting the pack. When they are closer to full sized, at least two years old and at least 140-150 pounds (64-68 kg) for a full-size breed, they can start learning to carrying a saddle with a light load—usually no more than 10% of their own weight. (A goat's weight can be closely estimated using a weight tape purchased from a goat supply.) You don't want to hurry their career, as it's important for their bones to mature as well as their minds. Once mature and in top shape, they should be able to carry up to 30% of their weight. So, a 200-pound (91 kg) wether (yes, they can get that big!) might carry up to 50 pounds (23 kg) of supplies.

If you have other, adult pack goats, there's great benefit in taking the younger, still-in-training animals out on hikes alongside the working goats. As long as the trip length and difficulty level do not exceed the young goats' abilities, they will benefit from the experience.

Pack and Cart Goats

Elyse Nicholson with Nigerian Dwarfs, Fairy Tale and Goose Girl testing out a new cart in 2013. Photo courtesy of Elyse Nicholson.

Equipment and Training the Cart Goat

Much of the training done with pack goats, with the exception of saddle training, applies to the cart goat. Even suggestions for choosing the right animal are almost identical. But there are some big differences. First, rather than follow you, the goat needs to learn to go ahead of you. They also need to learn to wear a bridle, and of course, to have a conveyance attached and following them everywhere when they are in harness.

You will need a harness and collar; a bridle, or driving halter; and, eventually, a cart. Start the training the same as for pack goats earlier. Once your goat has mastered being led, you can begin conditioning it to wear the bridle (a goat bridle, or driving halter, fits around the goat's head and has a ring on either side of their cheeks where the reins can be attached). At first, lead them around by the collar, but with the bridle on. Continue doing that, and gradually switch to using the bridle and reins instead of the collar. Done properly, the goat will barely notice the change.

Part IV   Pack, Cart, Brush, Pet and More

> **Off to the Races**
> You might stumble across a handful of online videos showing . Almost always, these are feral goats that have been corralled and harnessed. A dude climbs into the cart and then the goat is released. Of course the goat runs. Of course it looks energetic. But it's all because the animal is terrified. For the most part, this practice has been abandoned or modified, with trained goats being used instead. But, like with many things set to music and put on a video, you might be inclined to laugh. As with fainting goats, though, there's really not much funny about it.

Now comes the hard part in your cart goat's training: teaching the goat to go in front of you. Unlike pack training when you teach the goat to always stay behind you—and contrary to the goat's instinct of following a herd leader—the cart goat must move in front of you. The best way to do this is to have a second person working in front using a neck collar to encourage the goat forward, while you walk behind the goat. A training harness is useful at this time. Long reins are run from the bridle through side rings on the harness. As you walk behind the goat, give gentle guidance with the reins. Initially, these cues should be reinforced by the person going before or beside the goat. Reward the animal with praise, and perhaps a treat, to help them adjust to these new conditions. (Don't overdo it with the treats, or you'll simply train the animal to work only for such rewards!) Gradually, the person guiding the goat forward can diminish their cues while you increase yours. No matter what, try to make the experience pleasant and end on a positive note.

When the goat is ready, you can switch from the training harness, if it was in use, to the working harness. A working harness and collar distribute the weight of the cart's shafts over the goat's body as

Pack and Cart Goats

well as allows them to pull the conveyance. As with the rest of the animal's training, this is best done in stages. Before hooking them to a wheeled cart, it's a good idea to use a purchased or homemade wooden or aluminum frame that runs through the shafts and drags on the ground, making a travois. Be sure the goat is not afraid of this contraption before you attempt to hook it to its harness! Practice with the travois until the goat is comfortable. Then you're ready to graduate to the cart.

Some people use four-wheeled wagons and others two-wheeled carts. A proper goat cart that's light enough for a goat to pull is not cheap. A goat will only be able to pull about 1 ½ times their own weight! So, a lightweight cart with less drag—the resistance of the wheels on the ground—is ideal, particularly if you want to give more than a child a ride.

Farmer Profile: Brush Goats

Linda Williams, WestSide Goat Girl,

Gaston, Oregon

Linda and buddy, Napoleon. Image courtesy of Linda Williams.

In 2016 Linda Williams was looking for a change. The industrial engineer had just retired from Intel, and she and her wife, Norma, decided to give up urban life and move to the country. Very early in 2020, they found a twenty-acre property in the tiny town of Gaston, west of Portland, Oregon. Overrun with poison oak, Scotch broom, and blackberry, the property seemed in need of goats. After searching Craigslist for their goal of two to three goats, they stumbled on WestSide Goat Girl, a well-established goat grazing company, being sold by its second owners. The package deal included not just the twenty-six goats currently part of the herd, but portable electric fencing, a trailer, and, best of all, a customer list. As luck would have it, the original company name was aptly pertinent to both Linda and the location of her nascent farm in Gaston.

Linda calls her iteration of the brush-clearing business a "self-fashioned rescue." She keeps each goat until the end of its days, providing quite the interesting, abundant life until then. After all, as Linda points out, "Who wouldn't want to eat for a living?" Being in the Pacific Northwest, the wet weather favors the vigorous growth of the invasive plants goats adore, but it is not ideal for year-round goat grazing, what with goats being notorious for their dislike of rain as well as prone to parasite infection when grazing wet grasses. Linda says for six months out of the year the goats browse all their nutritional needs, but the other six she must feed them hay. The herd of thirty-six (Linda says she's currently at her maximum head count) performs about twenty grazing jobs per year, averaging one per week during the fair-weather months.

Linda is selective about the jobs WestSide Goat Girl accepts, meaning she's able to avoid some of the (Continued next pg.) (Continued) hazards facing goat shepherds offering similar services. For example, I asked her about trouble with pet dogs or humans bothering the goats when she's not there. Linda says one of her criteria is accepting jobs where no obvious hazards exist. In addition to checking on the herd each day, Linda posts signs at regular intervals around the electrified pen that educate the public about the goats, the danger of the electrified fence, and her own emergency contact number, just in case a goat escapes.

Escaping goats brought up the topic of training working brush goats to respect an electric fence. Linda encourages each goat to experience a mild—but memorable—shock from the fence early in their career. Goats are smart and retain the negative memory, avoiding the fence usually for the rest of their lives. At the job site, Linda's able to quickly set up an effective enclosure and move it as each area of the job is completed.

Unlike some goat browsing companies, Linda provides a "to the ground" service, following behind the goats with her own labor, cutting down saplings and shrubs too woody and coarse for the goats to process. Her website (see below) is quite thorough, offering examples of scenarios where her herd might be of use as well as pricing. Early on, one of her biggest issues was with prospective customers approaching her with the sentiment of *them* being the ones to do *her* a favor: providing what they saw as free feed. By sharing her pricing packages, she eliminates all but the most serious and appreciative jobs—many of whom are repeat customers.

Linda hopes to one day find a younger woman, whom she describes as a "mini-me," to take over the business, which she's shepherding toward nonprofit status with an admirable, long-term plan for succession.

To learn more about Linda, the goats, and WestSide Goat Girl, visit https://www.wsgoatgirl.com.

Pet, Therapy, and Yoga Goats

Chapter Sixteen

The Brush Goat

Using goats for organic brush and invasive plant remediation and wildfire fuel reduction has become quite popular in the last decade or so. Private land owners and government agencies at all levels are appreciating the efficient, low-impact abilities of goats to thin, clear, and control the growth of plants that are otherwise almost impossible, even with the use of the latest and greatest herbicides, to control. In this section we'll focus on the professional brush goat and brush goat business.

Given the chance, all goats are brush goat! The Pholia Farm herd out for one of our regular "goat walks" enjoying all sorts of nature's bounty—and helping keep the forest healthy.

Part IV Pack, Cart, Brush, Pet and More

Goats used for brush control might be any kind of goat, but large, hardy breeds are the most effective. They can reach higher into the canopy to browse, have larger mouths for eating thicker stems and larger thorns, and are at slightly less risk from predators. Brush goats might be adult wethers; meat or fiber goat brood does; culled dairy does that are healthy, but past their peak of milk production; younger meat goats that work as brush goats until heading to the finishing farm; or any combination of the above. If you're choosing goats for a brush business, it's important to understand the following: the nutrient requirements of the animals when compared to what the land might have to offer; fencing and shelter needs; supplemental feed; and predator protection. Additional business considerations include planning how to efficiently transport the herd; options for replacing and culling animals that don't work out; and assessing the market for your services.

Electric net fencing dividing a "before" area from an "after." Photo by Linda Williams, WestSide Goat Girl.

Pet, Therapy, and Yoga Goats

When I was working on the first iteration of this book back in 2015-2016, my editor sent me a link to an article concerning a brush goat business that was contracted to help eliminate invasive English ivy, *Hedera helix*, and other plants in an Oregon town. The goats were *too* effective, eating both the ivy and the trees the ivy was to be removed from. In addition, manure wasn't managed and possibly smelly buck goats were included in the group, creating poor public relations and not helping the image of goats. So, while a brush goat business might sound like an easy way to make… a buck, it must be approached in a conscientious way to increase the likelihood of success.

Equipment, Control, and Training

The equipment needed for working brush goats focuses on containing them safely; keeping them from plants that need to be protected; providing water, shelter, and supplemental feeding if needed; and transporting them to and from locations. All these issues must be considered first and addressed properly or your brush goat enterprise will shortly be out of work.

Fences

Good portable, electric net fencing, electrified to the right voltage, is essential. Portable solar chargers that keep a battery will likely need to be a part of the system—unless depleted batteries are to be swapped out by hand. Even more essential is training the goats to respect the fences BEFORE they are out on the job. This is best done at home when the goat is young or first brought in for the purpose of being a working brush goat. Different farmers do this conditioning in different ways, but in general you'll want to set up the electrified fence in a place where wet or damp earth provides the opportunity for the electricity to flow readily from the fence, through the animal, and back to the grounding rod (a long, conductive rod driven into the earth beside the charger/energizer). If the soil is too dry, then little or no current will flow from the fence and the animal might not feel a shock. I suggest placing the "trainer fence"

Part IV   Pack, Cart, Brush, Pet and More

inside an already fully fenced pen, that way the portable fence is not the only thing keeping the goats in should something go amiss. Allow, or even encourage, the new goat to touch its nose to the fence. Goats, being quite intelligent, will likely never forget that first shock!

If the job requires protecting saplings or trees with bark that's easily stripped by goats (who enjoy bark for both its nutrition and therapeutic benefits), you have basically three choices: you can turn down these jobs; you can alert the client to the issue, setting the expectation of loss and/or allowing them the chance to put wire, or other such barriers, around these plants; or you can include or charge a fee for this service. If your goats are well conditioned to the fence, you can use sections of non-electrified electric netting around groups of saplings or individual trees. Other options include using sections of rigid welded wire panels or encircling the trunks with hardware cloth (wire mesh) or small-hole chicken wire. (For more on electric fencing see *Holistic Goat Care*, pages 49-51.)

Water, Shelter, and Supplemental Feed

Water must be provided on-site, either from the client's supply or a tank brought in by the goatherd. Shelter is rarely used. Rather, goats browse during periods not requiring protection. A salt and mineral lick or feeder can be placed near the water. A baking-soda dispenser is also recommended so that any feed changes and rumen acidosis that might result can be dealt with by the goats. Supplemental feed should only be needed if the area has been cleared of foodstuffs but the herd cannot yet be moved. In that case, extra feed is critical to keep them from challenging the fence in an effort to reach more food! Professionals such as Linda Williams check their herd regularly and stay on top of such needs before they become a problem.

Transporting and Handling

Loading goats in a trailer is not nearly the training challenge that loading horses is! In fact, usually a bit of food and one goat being led in is all it takes. Still, the goats should be comfortable with humans so that they are easy to load and unload at different sites. This will also help on-site as

Pet, Therapy, and Yoga Goats

they won't be frightened by the inevitable spectators and possible innocent/ignorant harassment by people if the site is public. Signs should be made and posted that warn people about the fence and give guidelines about interacting with the goats or leaving them alone—including instructions not to feed them. Include your contact number on the signs in case something untoward occurs, and as a way of advertising.

Training Goats to Browse

Goats aren't born knowing the plants they should and should not eat. Some won't even like edible varieties the first time they try them! Most goats learn what to eat by watching their mothers or peers enthusiastically noshing on things such as grape vines and poison ivy/oak and avoiding things that are toxic. Goats that don't have this type of role model will need a little help from you. When we first moved our Pholia Farm herd to the Oregon property, with all of its poison oak and blackberry, the goats wouldn't touch it. So with a little tough love, I deprived them of their regular dinner a few nights in a row and instead filled the feeder with branches of lush poison oak. It didn't take long for them to start nibbling, and after a few days they relished it, both in the feeder and in the forest.

Goats making short work of invasive English ivy.
Image courtesy of Linda Williams.

Part IV   Pack, Cart, Brush, Pet and More

Poisonous and/or toxic plants need to be considered. You as the goatherd need to be able to preview the client's land and discover if there are any seriously toxic plants. Then it's up to you to decide the risk and either turn down the job or remove the risk. Keep in mind that goats are very selective and have a protective memory. A plant nibbled once, if followed by a tummy ache or other untoward symptoms, is rarely tasted again. In addition, many plants toxic to cattle and sheep are not harmful to goats; this means it's difficult to find accurate information about just what is dangerous. For more on poisonous plants, see *Holistic Goat Care*, page 97.

Left to Right – Angel, Bam Bam and Lumpy, a trio of working goats. Image courtesy of Linda Williams.

Breeder Profile
Amanda and Scott Peterson
Oats and Ivy Farm, Somis, California

Amanda and her full-plank "helper."

During our almost twenty years of running Pholia Farm we had the great fortune to work with some awesome interns through the WWOOF program (World Wide Opportunities on Organic Farms). Of that group, several, including Scott Peterson and Amanda Nunez, remain treasured friends. Trained as chefs, the couple came to our farm in 2015 determined to gain the skills needed before starting their own small goat dairy on family land in Somis, in Southern California. Like many small towns in the valleys bordering the Los Angeles basin, Somis is a mix of suburban-type developments and agriculture. Although the pair's dream was modified somewhat, they've succeeded in building a sustainable small goat farm, thanks, in great part, to fulfilling a need—not for goat cheese, but for goat's milk soaps, and, perhaps even more importantly, goat "snuggles."

With a fluctuating population of between thirty-five and sixty goats, fifteen being bred and milked, Amanda and Scott keep very busy on their multigenerational ten-acre lot. They sell their soaps and candles at several area farmers' markets, to farm guests, and through mail order. When I visited, they were hard at work filling a custom order for a wedding, breeding does for the spring, building a house (themselves), and caring for not just their ambitious preschooler, but a newborn. As if that isn't enough, they also run an Airbnb that attracts many guests wanting an on-farm experience close to the big city. As it was autumn when I visited, their other income source, "goat snuggles," was on break.

Before snuggles, Oats and Ivy offered goat yoga. When the farm first started, goat yoga was just becoming a craze. In first considering the option, Amanda wrestled with whether they could offer it without feeling as if the goats were being exploited. They decided that as long as they weren't breeding goats just to have cute babies for the classes, they would be fine. To further that perspective, the farm often included the mom goats in the yoga sessions, entertaining people with the sweet interactions of the moms and kids. Initially, Oats and Ivy was one of the only places in the region offering goat yoga, and classes were packed! They charged $35 a person, splitting the proceeds 60/40 with the guest yoga instructor. For the first four years, classes filled to the maximum limit of twenty-five people, but then the goat yoga market became saturated, and attendance dropped to just a few. It was then, Amanda says, that she thought about the dynamics of their classes and realized people weren't necessarily coming for the yoga, but to be around friendly, cuddly goats. "It was a no-brainer," she says about deciding to offer sessions sans yoga. These snuggle sessions turned out to be a win in more than one way.

"By switching from goat yoga to goat snuggles, it opened up time with the animals to everyone—not just people who could or wanted to do yoga. Half of what people come for, other than the ridiculous cuteness of the goats, is the story of our little farm." During the spring and early summer, sessions are offered three times every weekend. Guests pay $25 a person for a ninety-minute session to sit on straw bales, groom and snuggle the goats, and learn about farm life—in micro detail. "We get lots of repeat guests," Amanda says. "People get to know the goats as individuals. They are invested."

And it's not just city folks with expendable income that come to the farm: Oats and Ivy hosts groups of high-risk kids in a mental health program that brings them to the farm. Nunez says, "Most arrive looking closed-off and shy, but everyone leaves with smiles on their faces."

Whether you want to add agritourism to your farm income or are simply looking for a reason to have goats, their value beyond milk, meat, and other goods is very real. To learn more about Oats and Ivy Farm and Amanda and Scott, visit https://oatsandivyfarm.com.

Chapter Seventeen

Pet, Therapy, and Yoga Goats

Although I've put this section last, it's really one that stretches across all categories of both the usefulness of goats and why so many of us love them. In this section we'll focus on goats whose primary purpose is companionship—in one form or another. I mentioned in the introduction that my first goats were companions for my horses. First there was Sundance, an Alpine-pygmy cross, then Amigo, a pygmy, then Goatee Goat, also a pygmy. They kept my horses company, but were equally enjoyable for me to have around.

Amelia Caldwell (daughter) in 2011 with one of her beloved dairy goats, Baklava.

Pet goats are found on many urban/suburban and rural small lots as well as on large and small farms. Some city pet goats are also milk

Part IV   Pack, Cart, Brush, Pet and More

and/or fiber goats, providing companionship, tasty milk, and useful fiber. Most pet goats are small breeds such as Pygmy and Nigerian Dwarf. Some are retired brood and/or milking does, but the vast majority are castrated males that otherwise might have ended at—what we goat folks call—freezer camp.

Author of the memoir Farm City, Novella Carpenter, caught the attention of many in the early 2000s with her urban Oakland, California backporch goats.

Pet Goat Considerations

The legality of keeping goats on small lots varies widely. In recent years more and more major cities have allowed property owners to house a small number of (usually) small goats. Before you bring home your next furry friend, be sure to know what the regulations are in your area. An internet search such as "MY CITY zoning livestock regulations" should yield the right links. Be sure to include the state your city is in, as there are quite a few duplicate city names across the US. (You might end up in a situation similar to the one I was in when I booked a hotel room in Portland [Oregon, I thought] only to arrive to find out my reservation was for the same chain hotel in Portland, Maine.)

Training Pet Goats

Pet goats should receive the same initial training as those that will be used for packing or cart training. They need to be bonded to humans but also taught manners—just like a well-behaved pet dog should be trained to not jump on people and to walk on a leash. They should also be trained not to butt people. This is more a matter of NOT training them *to* butt! Roughhousing with goats encourages this natural but potentially harmful behavior. Also, people tend to naturally want to rub a goat on its forehead, but that should be avoided, since it usually stimulates their desire to butt and rub.

Therapy Goats: An Untapped Opportunity

The term "therapy animal" is used to designate an animal screened for their ability to comfort people in institutional settings or under such care. Examples are nursing homes, schools, and penitentiaries. Such goats can provide much-needed interaction for folks dealing with a variety of isolating issues. Whether dogs or goats, therapy animals must have the right personality: calm and tractable, in addition to the ability to be trained as described above. In addition to training, therapy animals will need to be cleaner than normal (baths, which goats instinctively dislike, might be necessary), and a way to contain manure and urine (usually in the form of a dog diaper) must be addressed.

If you're interested in doing something like this with your pet goat, I suggest a web search first for "animal therapy programs near me." The program may not currently include goats, but that doesn't mean they won't consider it.

Part IV   Pack, Cart, Brush, Pet and More

There's no need to train baby goats to jump and frolic for a yoga class! Image courtesy of Amanda Nunez

Goat Yoga: Forget the "Down Dog"

This section title is basically an oxymoron. Other than not wanting aggressive goats, goat yoga relies on gentle goats running amok in the midst of a group of people doing downward-dog (*adho mukha svanasana*) pose. Goats that have been hand-raised (usually bottle-fed) are naturally drawn to people. And the younger the goat, the more likely it will want to jump on a person's back—it's just what they do!

Special Health Needs of Pet, Therapy, and Yoga Goats

Pet goats have similar health needs to those of working goats. Although their lives might not be as stressful, they can suffer from too much love. For example, there's often the temptation to allow a pet goat to spend a good deal of time living indoors. This restricts them from the sunshine and exposure to microbes that benefit their health. In addition, pet goats are quite often overfed, OR fed treats and feeds that are not balanced for their nutritional needs.

Pet, Therapy, and Yoga Goats

If you're considering a pet goat, you owe it to the relationship to understand goats 'nutritional and health needs based on gender, age, and job (whether they're producing milk or just lazing around greatly influences those needs). Wethers, the most common pet goats, have very special needs due to a high risk for developing deadly urinary calculi (see *Holistic Goat Care*," urinary stones," for more). Their diet must be carefully balanced in order to prevent stone formation. Like a dog, a pet goat is a commitment that deserves a lot of forethought for you and your goats to be happy for the long haul.

Finding a vet to help care for your goats is often one of the biggest challenges facing even rural farmers. Please be sure to reach out to area vets BEFORE you need one! Finding other goat lovers in your area, whether through a social media platform or at the feed store, can be very helpful to the important quest of finding a veterinarian that will take you, and your goats, on as clients.

Goat milk soaps are another terrific way to optimize goat milk. These bars are made by Oats and Ivy Farm. See profile page 177-178.

Part IV Pack, Cart, Brush, Pet and More

End of Life

You may not want to think about this just yet (and it's a heck of a place to end the book), but keep in mind that someday your goat will reach the end of its life. This will bring new decisions. First, you should have a vet willing to perform euthanasia, should it become the humane option, or, if you are a farmer or hunter, be prepared to perform this service yourself. I write quite a bit about this in *Holistic Goat Care*, and, of course, it's discussed earlier in this book in Chapter 8. However, you decide to stop the suffering of your sweet pet, it will of course be difficult, but it will be even more so if you don't have that option figured out ahead of time.

You will also have to deal with your pet goat's remains. Veterinary offices can, of course, dispose of them for you, and in many areas you can deliver animal carcasses to landfills if you have no other option. If you are on a small lot, it's unlikely you will legally be allowed to bury your goat—or other pets—on the land. Some folks choose cremation, a service available to most small-animal owners. Others have in mind a rural friend with a pet burial area you can visit. There are even green burial and conventional cemeteries offering plots for beloved pets if your budget and desires allow.

End Note

Being a dairy cow and horse person, I never thought goats would take over my life in the myriad ways they have. I have the utmost respect for the little rascals. May your journey with caprines bring the same fulfillment and joy. Let me leave you now with the profile of a wonderful man, Daniel Laney, and his mission to help goats and people in the faraway land of Nepal.

Daniel Laney
Worldwide Goat Project, Nepal

Daniel arriving in Nepal with a load of supplies.

Goat folk are often as interesting as goats themselves. Possibly no one exemplifies this as well as Daniel Lee Laney and his work for the benefit of the people and goats in Nepal. The Worldwide Goat Project Nepal (WGPN) began in 2013 when Daniel sought to escape a dark segment of his own life following a terrible illness, a monthlong coma, and then the death of his mother. Already a world traveler especially fond of the land and people of Nepal, and a goat expert (having raised dairy goats since 1972, having been an American Dairy Goat (ADGA) senior judge, and having been the past president of ADGA) the mission was a natural fit.

His original idea was to bring improved genetics to the farmers of Nepal, where goats are a main source of nutrition and therefore survival, through the import of Kiko semen (see pages 57 and 61 for more on Kiko goats), but in the process of working with government officials toward this goal, Daniel learned of a herd of fairly recently imported Saanen goats that were not thriving. Intrigued, he made his way, via a long, winding trip on bus and motorcycle, to the National Goat Research Station (GRS) perched high in the mountains (3,576 ft, 1,090 m) in the idyllic community of Bandipur. There he found researchers used to working with the areas many hardy breeds, but not with the (Continued next pg.) (Continued) focused, high-production genetics of the imported Saanens. In addition, because these new goats were so expensive to bring in (costing, when all expenses were added up, about $2,000 each) they were being sheltered as if, Daniel says, "they were made of gold." In other words, in a fashion not conducive to goat health and productivity.

Slowly, through getting his hands dirty by working side-by-side with the caregivers, he was able to integrate his expertise with GRS's goals, leading to better care for the goats, while always being conscious of respecting native and local wisdom and traditions. The health of the imported herd has improved and inroads have been made in the care of native breeds as well, primarily through steps such as ensuring ample water for the goats, planting forage/browse trees that can be used as a foodstuffs supply, and teaching the value of avoiding inbreeding in herds. To that end he predicts a "very encouraging sustained future of Saanen and Saanen-cross goats in Nepal as more Nepali farmers see the value of having the dual-purpose crosses for both meat and milk." He's also facilitated a slow but steady growth in the goat cheese market, increasing options for farmers to better their lives.

Laney's efforts aren't limited to providing advice and elbow grease. He also collects equipment and supplies in the United States and personally ferries it all the way to Nepal. How, you might ask, is this funded? Much of it comes from his personal contribution, but goat groups throughout the nation are also pitching in. In fact, I was at a local goat meeting just a few months ago here in my corner of Oregon and they just happened to be voting on a donation to WGPN. There's a Facebook group, maintained by Daniel, with updates on his adventures and photographs galore that make you want to ask him if he might need an extra set of hands on his next trip.

Daniel has encountered many unique breeds during his time in Nepal, including the Sinhal, a beautiful meat and fiber goat with a long flowing coat found in the higher (Continued next pg.) (Continued) elevations; the

Khari, formerly known as the "hill goat," a hearty meat breed that can raise three crops of kids every two years; and the breed he'd most like to bring home, the Barberi. Although not native to Nepal, they are flashily colored, have a variety of ear types, and are small enough, Daniel says, "to fit in a suitcase." But no matter the breed, he says," Seeing goats in other countries dealing with very diverse climates and management practices, exhibiting unique breed phenotypes, and yet still thriving, exemplifies just how adaptable they are. It's only intensified my appreciation for our amazing goat!"

I asked him what he sees for the future of Worldwide Goat Project Nepal. He answered, "I would like this project to become a 501c3 nonprofit so we can expand aid such as water filtration cubes into more village schools and afford a semen tank for AI purposes. I am currently working on that goal. Hopefully this next year will see that become a reality. I can also," he added at the end of our conversation, "see a younger person stepping into my shoes as my own hiking boots get worn a bit thin." To learn more, visit https://worldwidegoatprojectnepal.org.

Barberi goats, a flashy, multi-purpose breed that Daniel Laney is particularly drawn to. Photo taken in Nepal by Daniel Laney.

Resources

General Goat Resources

Books

A Holistic Vet's Prescription for a Healthy Herd, by Richard Holliday, DVM and Jim Helfter

Goat Medicine, by Mary C. Smith and David M. Sherman

Holistic Goat Care, by Gianaclis Caldwell

The Mysterious Goat, by Dr. C. Naaktgeboren

Natural Goat Care, by Pat Coleby

Raising Goats Naturally, by Deborah Niemann

Forums, Websites, and Groups

The Goat Spot Forum, https://www.thegoatspot.net/

Goat Depot. Lots of great articles and information about goats around the world. https://goatdepot.com

Oklahoma State University, Breeds of Goats, https://breeds.okstate.edu/goats/

Supplies

Caprine Supply, https://www.caprinesupply.com/

Hoegger Supply, https://hoeggerfarmyard.com

Premier1, https://www.premier1supplies.com

Sydell Sheep and Goat Equipment, https://sydell.com

Farm Tek, https://www.farmtek.com/home

Dairy Goat Resources

Associations

American Dairy Goat Association, https://adga.org/

American Goat Society, https://americangoatsociety.com/

Miniature Dairy Goat Association: https://miniaturedairygoats.net

Nubians: International Nubian Breeders Association, https://www.inba-nubians.com/

Lamanchas, https://www.lamanchas.org/

French and American Alpines: Alpines International, http://alpinesinternationalclub.com

Oberhalsli: Oberhasli Breeders of America, https://www.oberhaslibreeders.com

Saanen, No breed club at the time of publication.

Sable Saanen: International Sable Breeders Association, https://sabledairygoats.com/about-sables/

Guernsey Goats: Guernsey Goat Breeders of America, https://www.ggboa.org/

Nigerian Dwarfs: The American Nigerian Dwarf Dairy Association, https://www.andda.org/

Books, Videos, Articles

The Dairy Goat Production Handbook, Merkel, Gipson, Sahlu, Langston University Press

The Small-Scale Dairy, by Gianaclis Caldwell

Storey's Guide to Raising Dairy Goats

Supplies

Hamby Dairy Supply, https://hambydairysupply.com/

Meat Goat Resources

Associations and Sites

American Boer Goat Association, https://www.abga.org/

American Meat Goat Registry, https://www.amgr.org/

Meat Goat Society, https://meatgoatsociety.com/

Books, Videos, Articles

Basic Livestock Handling, by Temple Grandin

Butchering: Poultry, Rabbit, Lamb, Goat, Pork, by Adam Danforth

The Meat Goat Handbook, Langston University (the 3rd edition will be out mid-2025; until then, the book is hard to find)

Meat Goat Selection Carcass Evaluation and Fabrication Guide, Louisiana State University, https://store.lsuagcenter.com/p-64-meat-goat-selection-carcass-evaluation-fabrication-guide.aspx

Storey's Guide to Raising Meat Goats, Storey Publishing

Cornell University Sheep and Goat Marketing:
http://www.sheepgoatmarketing.info/#&panel1-1

Langston University Meat Goat Certification,
https://certification.goats.langston.edu/mdl/course/index.php?categoryid=7

USDA meat inspection program, https://www.fsis.usda.gov/inspection/

USDA Institutional Meat Purchas Specifications,
https://www.ams.usda.gov/grades-standards/imps

Fiber Goat Resources

Associations

American Angora Breeders Association, https://www.aagba.org/

Cashmere Goat Association, https://cashmeregoatassociation.org/

Pygora Breeders Association, https://www.pygoragoats.org/

American Nigora Goat Breeders Association, http://www.angba.org/

Books, Videos, Articles

Book: *Angora Goats the Northern Way*, by Susan Black Drummond

Book: *Raising Angora Goats for Beautiful Mohair*, by Sharon Chestnutt

Book: *Raising the Angora Goat for Profit*, by William Black

Video: "Legendary Kashmir Goats Cashmir processing," YouTube, https://youtu.be/NbE7i25a580?feature=shared

Book: *Raising Animals for Fiber: Producing Wool from Sheep, Goats, Alpacas, and Rabbits in Your Backyard*, by Chris McLaughlin

Hide Tanning Resources

Books, Videos, Articles

International Leather Club article on brain tanning: https://www.internationalleatherclub.com/brain-tanning/

Natural tanning articles and classes, https://braintan.com/intro/intro.html

How to smoke a hide, https://anchoredoutdoors.com/the-hows-and-whys-of-smoking-your-pelts-and-hides/

Skill Cult blog and YouTube channel with excellent information on all types of natural tanning. It might take a little searching to find the videos and posts you want, but they're worth it.

Blog: https://skillcult.com/blog/tanningmaterials

YouTube: https://www.youtube.com/c/SkillCult

Supplies

The Tannery Inc., https://thetanneryinc.com/

Fur Harvester's Trading Post, https://fntpost.com/

Cart and Pack Resources

Associations

The North American Packgoat Association, https://napga.org/

Books, Videos, Articles

Articles: Working Goats, https://workinggoats.com/

Supplies

Quality Llama Products Inc., https://llamaproducts.com/

Pack goat equipment, https://packgoats.com/

K-9 Carting, https://k9carting.com/

Index

A

Aging, 76, 77

Al and Lin Schwider, 91

alum, 125, 138

Amanda Nunez, 177

American Dairy Goat Association, 2, 13, 15, 19, 23, 190

Anglo-Nubian. See Nubian

angora, 101

Angora, 60, 91, 101, 103, 106, 107, 108, 110, 111, 112, 116, 118, 119, 192

Angoras. See Angora

Australia, 60, 84, 110

B

Bark Tea, 142

Body Capacity, 7

body condition, 65, 66

Boer, 39, 56, 59, 60, 61, 191

Boki goats, 61

brain tanning, 126, 145, 192

Brine Aging, 77

Brush Goat, 171

bucket-washer, 35

Bucking. See Dehairing

Buckskin, 128, 132

butcher, 64, 66, 71, 78, 85, 145

Butchering, 65, 70, 77, 78, 191

C

cabrito, 54

captive bolt, 68, 69

Captive bolt pistol, 66

cashmere, 60, 101, 103, 104, 106, 109, 110, 111, 112, 116, 117

Charles McCartney, 155

Cheese, 38, 40

chevon, 54

Chevon, 89

chevre, 86

Chevre, 46

chrome tanning, 125

Combing, 104

conformation, 2, 26, 156

Cooler Shrinkage, 65

cooling milk, 36

Cornell University, 84, 191

D

dairy clip, 32

Dairy Herd Improvement, 24

Dairy Strength, 5

Daniel Laney, 184

Dehairing, 104, 132, 133
Dr. Grandin. *See* Dr. Temple Grandin
Dr. Temple Grandin, 68
Dressing Percentage, 66
dry, dark, and firm, 67
dry, firm meat, 67

E

electric net fencing,, 173
extensive management, 61

F

Fat (Brain) Tanning, 126
Felting, 104
firearm, 66, 69, 70
fleece, 103, 104, 105, 106, 107, 112, 113, 116, 118, 119
Fleece, 106, 112
Fleshing, 128, 130
Freezer burn, 80
full-grain, 129

G

Genemaster, 61
general appearance, 2, 3, 5, 8, 55, 56, 106, 107, 156
General appearance, 3
Goat Man" of Georgia, 155
Goat meat, 53, 54, 94, 95
goat roast, 93

Goat's milk, 36, 39
Golden Guernsey, 20, See Guernsey
Guernsey, 13, 20, 190
Gutting, 74, 75

H

HACCP, 83
halal, 57
Halal, 68
Hanging Weight, 66
herdbooks, 13, 23

J

Jerky, 95

K

Kefir, 41, 50
Kiko, 53, 61
Kinder, 21, 61

L

lactose, 40
Lactose, 21
Lamancha, 16, 39
Langston University, 54, 83, 85, 190, 191
Linda Williams, 168
LindLinda Fox, 120
Live Weight, 66
Lookout Point Ranch, 57

M

mammary system, 2, 7, 8, 9, 10, 55

Mammary System, 7

Marc Warnke, 160

Meat-to-Bone Ratio, 66

medial suspensory ligament. *See* suspensory ligament

micron, 102

Milk fat, 21

milk fat,, 25

milking machine, 32, 34, 35, 36

milking star, 25

Mineral tanning, 138

Mineral Tanning, 125

mobile farm killer, 70

mohair, 101, 102, 103, 106, 107, 110, 111, 117, 118, 119

Mountain Lodge Farm, 62

Mutton, 54, 86

N

Nigerian Dwarf, 10, 19, 39, 112, 180, 190

Nigerian Dwarfs, 8, 13, 20, 23, 103, 190

Novella Carpenter, 180

Nubian, 17, 21, 39, 60, 190

O

Oberhasli, 12, 13, 15, 39, 190

Oiling, 140

On-farm kills, 67

P

Pack and cart goats, 156

Packsaddle, 159

Paneer, 40, 43

pannier bags, 157

Panniers, 158

Parr. *See* Tami Parr

pedigree, 14, 23, 24, 25

pH meter, 135, 136

Pickling, 129, 135, 136

Primal Cuts, 66

pulsator, 34

Pygora, 112, 113, 114, 192

R

racing cart goats, 166

Rawhide, 129

Richard Johnson, 57

Robin Oliver, 114

Rovings, 105

S

Saanen, 8, 13, 14, 15, 16, 39, 190

Sable, 13, 15, 190

Sausage, 97

Savanna goats, 61

shearing, 103, 106, 113, 117, 118, 119

Sherwin Ferguson, 62

Shrinkage, 66, 105

Skinning, 72

Slaughter, 65, 67

Softening, 140, 144

Spanish, 16, 54, 56, 60, 61, 110, 111

Split-leather, 129

strip cup, 33, 35

Suede, 129

Superior Genetics, 26

suspensory ligament, 10

Swiss Alpines. *See* Alpine

T

Tallow, 98

Tami XE "Parr" \t "*See* Tami Parr" Parr, iii

Tami Parr, 12

Tami Parr, 86

Tanning, 125, 127, 128, 129, 137, 142, 145, 147, 192

Tawing, 125, 137

Tennessee fainting goats, 61

The American Goat Society, 23

The Mysterious Goat, 155, 189

Therapy Goats, 181

Toggenburg, 6, 12, 13, 14, 15, 39

U

USDA, iv, 24, 59, 68, 82, 84, 86, 87, 119, 191

V

Vegetable Taning, 141

Vegetable Tanning, 125

Vitamin B, 21

W

Wood Ash, 133

Y

Yoga Goat, 182

Let's Connect

For discounts, news of new books, book club info, or just to connect, find me at the following:

 [Subscribe to newsletter](#)

 https://gianacliscaldwell.com

 gianaclis@gmail.com

 @gianaclis.bsky.social

 https://linktr.ee/gianaclis

If you liked *The Useful Goat*, or any of my other books, please consider **writing a review or rating** the book on the platform of your choice.

Thanks, it really helps!

About the Author

Gianaclis (gee-on-a-klees) is a multi-award-winning author raised in the wilds of Oregon believing trees were spirits and failing at picking string beans. She writes cheesy (no really) non-fiction on cheesemaking and goats, and hopeful speculative fiction. In her spare time, she guides other old ladies in plies, ronde jambes, and the occasional pirouette. She still talks to trees; sometimes, they answer.

www.ingramcontent.com/pod-product-compliance
Lightning Source LLC
Chambersburg PA
CBHW062109290426
44110CB00023B/2756